Geopolitics

The term 'geopolitics' came into use at the end of the nineteenth century. Thinking globally was then formally connected by geopolitical reason to acting globally, but the actual practices of geopolitics began much earlier, when Europeans first encountered the rest of the world.

Geopolitics identifies and scrutinizes the central features of geopolitics from the sixteenth century to the present, paying close attention to its persisting conceptual underpinnings, novel turns and shifting impacts. The book focuses on four key concepts of the modern geographical imagination: visualizing the world as a whole; the definition of geographical areas as advanced or primitive; the notion of the state being the highest form of political organization; and the pursuit of primacy by competing states. The second edition is thoroughly revised to take into account recent world events and what they augur for the future of the modern geopolitical imagination, including the possibility that understanding of the geography of world politics – and a new world politics – can be rebuilt on alternative grounds.

Exemplified by topical issues such as the US 'war on terrorism', the rise of China as a Great Power, the break-up of the former Yugoslavia, US President G. W. Bush's identification of a global 'axis of evil', and the re-emergence of Central Europe as a critical geopolitical region, the book shows how questions of the organization of power combine with those of geographical definition and highlights the crucial geopolitical certainties from as recently as fifteen years ago which have now either disappeared or are in question. *Geopolitics* provides an invaluable introduction to current critical debates over geopolitics and world politics.

John Agnew is Professor of Geography at the University of California, Los Angeles.

PRAISE FOR THE FIRST EDITION

'John Agnew has long been recognized as one of geography's most creative theorists of the interaction between place, politics, and the economy. . . . In this book, Agnew is concerned with why the modern geopolitical imagination has had such a long-lasting hold on how the international political order is conceptualized. In other words, how and why did what eventually evolved into the standard principles of contemporary geopolitics, develop? . . . The book is well written, thoughtful, creative, and persuasive. It should prove to be an important milestone in the development of geographic thought in years to come.'

Fred M. Shelley in *Annals of the Association of American Geographers* (2000)

'. . . [I]t is an indication of the wealth of ideas introduced in Agnew's study that each chapter leads on to a detailed discussion of key elements of today's debates around politics, power and spatiality. The text is clearly written, coherently structured and attractively produced. *Geopolitics* deserves to be widely read by students of geography, politics and international relations.'

David Slater in *Progress in Human Geography* (2000)

PRAISE FOR THE SECOND EDITION

'At a time when "geopolitics" is so often at the center of the news and academic debate, this book provides a lucid historical and geographical study of the evolution of the modern geopolitical imagination. The book's intellectual breadth spans from post-colonial critique of Western-centric global mappings to international political-economy analysis of global finance. Through it all, the diverse ways in which visions of the "world-as-a-whole" have been deployed by powerful elites are carefully charted. The result is an exemplary geo-history of geopolitics and, as such, an important caution against arrogant uses of the term in popular commentary.'

Matthew Sparke, Associate Professor in Geography and the Jackson School of International Studies at the University of Washington, Seattle

'Concise and clear, historical and critical, this magisterial overview clearly shows how modern spatial conceptions of the world underwrite politics, economics, states, identities and violence.'

Simon Dalby, Ph.D. Professor and Chair, Department of Geography and Environmental Studies Carleton University, Ottawa Ontario

'In a world of proliferating references to geopolitics, the "re-visionings" in this accessible book explore how and why there is so much more to geopolitics than first meets the eye.'

James D. Sidaway, National University of Singapore

Geopolitics

Re-visioning world politics

Second edition

John Agnew

Routledge
Taylor & Francis Group

LONDON AND NEW YORK

First published 1998
by Routledge
2 Park Square, Milton Park, Abingdon, Oxon, OX14 4RN

Simultaneously published in the USA and Canada
by Routledge
270 Madsion Ave, New York, NY 10016

Second edition 2003

Reprinted 2006

Routledge is an imprint of the Taylor & Francis Group

© 1998, 2003 John Agnew

Typeset in Times and Franklin Gothic by
Florence Production Ltd, Stoodleigh, Devon
Printed and bound in Great Britain by
T J International Ltd, Padstow, Cornwall

British Library Cataloguing in Publication Data
A catalogue record for this book is available from the British Library

Library of Congress Cataloging in Publication Data
Agnew, John A.
  Geopolitics: re-visioning world politics / John Agnew. – 2nd ed.
    p. cm.
  Includes bibliographical references and index.
  1. Geopolitics. I. Title.
JC319.A43 2003
320.1′2–dc21   2003002691

ISBN 0–415–31006–7 (hbk)
ISBN 0–415–31007–5 (pbk)

Contents

List of illustrations   vii

Acknowledgements   ix

Chapter 1   Introduction   1

Chapter 2   Visualizing global space   15

Chapter 3   Turning time into space   35

Chapter 4   A world of territorial states   51

Chapter 5   Pursuing primacy   67

Chapter 6   The three ages of geopolitics   85

Chapter 7   A new age of 'global' geopolitics?   115

Chapter 8   Conclusion   127

Glossary   135

Bibliography   139

Index   147

Illustrations

Plates

2.1	A German woodcut of *c.* 1505 showing a representation of the Topinambas of coastal Brazil	20
2.2	The Crowned Britannia	27
6.1	Europe as a queen	89
6.2	'Indians from Canada make a sacrifice to the Great Spirit'	91
6.3	'Jungles to-day are gold mines to-morrow'	94
6.4	'Behind the enemy powers – the Jew': Nazi racial propaganda	97
6.5	Dismantling the Berlin Wall	107
7.1	'You really think this will work?'	121

Figures

1.1	A spectre haunts Central Europe	3
1.2	President George W. Bush's 'axis of evil'	4
2.1	Ptolemy's world outline, from the Rome edition of 1409	18
2.2	The world as known to Europeans, *c.* 1800	25
2.3	Mackinder's geopolitical model (1904)	29
3.1	Aboriginal land claims in Australia (1997)	38
3.2	The political geography of Italian unification, 1860–70	43
5.1	The Prisoner's Dilemma game	81
7.1	The world during (A) and after (B) the Cold War	117
7.2	Huntington's civilizations (1993)	119
7.3	Defence spending by country, absolute (A) and relative (B), 2000	123
8.1	Spatialities of power	130
8.2	Diachronics of spatialities of power	132

Tables

4.1	Two dimensions of state power and the four ideal types of state they define	55
4.2	Pillars of the emerging 'market-access' world economy	60
5.1	Conventional and ownership-based trade balances, United States (1986) and Japan (1983)	79

Acknowledgements

I would like to thank Nick Entrikin, Linda McDowell, Carol Medlicott, Felicity Nussbaum and Gerard Toal for helpful comments and suggestions on the first edition. Howard Harrington and Chase Langford gave timely assistance in producing the illustrations. Derek Gregory read the original manuscript with great care. His suggestions for rewriting were pivotal in making this book both more coherent and more accessible. Mathew Coleman, Vilho Harle, Sami Moisio and David Slater offered useful suggestions for the second edition. I, however, remain solely responsible for what I have made of all comments and suggestions.

The following gave permission to reproduce copyrighted material:

Plate 2.1   The New York Public Library
Plate 2.2   The British Library, London
Plate 6.2   The Getty Research Institute, Los Angeles
Plate 6.4   Bildarchiv Preussischer Kulturbesitz, Berlin
Plate 7.1   Cartoonbank.com for the *New Yorker*

1 Introduction

In the aftermath of the attacks on the World Trade Center in New York and the Pentagon (US Defense Department) outside Washington, DC, on 11 September 2001, US President George W. Bush declared war on 'terrorism'. This, it has become clear, involves much more than simply going after those who perpetrated the attacks – the so-called *al-Qaeda* network of Islamic terrorists led by the Saudi Arabian millionaire Osama bin Laden. The result has been nothing less than a total reorganization of world politics, with the US government claiming the right to intervene militarily wherever and whenever it wants, initially in Iraq but without precisely defined geographical limits. Rooting out terrorism all over the world has now replaced older themes in the US foreign policy lexicon, working backwards from most to least recent, of 'containing Communism' (in the former Soviet Union and its sphere of influence), creating 'collective security' after the First World War, or helping to maintain the balance of power between the European Great Powers in the late nineteenth century. The US government is now able to do more or less as it pleases (at least militarily) because it has become the single superpower. Of course, it also sustained the attacks on its soil because of its global geopolitical centrality and support for governments – particularly those of Israel and Saudi Arabia – that excite much hostility from Muslim extremists.

What is new after 11 September 2001 is that the world has never before had just one major power combined with a violent opposition from a shadowy network without a single state to its name (although it came close to having one in Taliban-dominated Afghanistan between 1996 and 2001). What is more familiar, however, is that the US government and its new adversaries are participants in a global 'game' for political control and influence that has dire consequences for the world as a whole. The game goes back to the creation of modern world politics beginning in the sixteenth century but getting into full swing only in the nineteenth century. It is the fruit of a modern geopolitical imagination that encompasses and directs the theories and practices of world politics. This geopolitical imagination has long framed world politics in terms of an overarching global context in which states vie for power outside their boundaries, gain control (formally and informally) over less modern regions (and their resources) and overtake other major states in a worldwide pursuit of global primacy. It is the combination of all these features that makes the geopolitical imagination a peculiarly modern one.

The period since the ending of the Cold War in 1989–91 has seen a number of other dramatic changes that, along with worldwide terrorist networks, might seem to challenge the continuing utility of the modern geopolitical imagination as a singular guide to practice in world politics. These would include the deepening of the European Union with the creation of the new currency, the euro, thus giving rise to a new form of polity that is neither a gigantic territorial state nor a simple common market; the collapse or decreased power of central government institutions in many states across the world – from Colombia in South America to Somalia in Africa and Afghanistan in Asia; the threat of national economic bankruptcy in the case of states such as Argentina, Brazil and Turkey; the tremendous growth of China as an export-based manufacturing economy supplying seemingly endless amounts of consumer goods to deindustrializing economies such as the United States; and the inability of the main parties to many nationalist conflicts, such as that between Israel and Palestine, to reach agreements over their mutual boundaries because each claims much the same territory as the other. The state-based world in which the modern geopolitical imagination has developed and to which it applies might seem to be in serious disarray.

The modern geopolitical imagination

The purpose of this book, however, is to identify the major elements of this long-dominant approach to world politics – the modern geopolitical imagination – and to show their historical–geographical specificity in the European and, later, American encounters with the world. In other words, the aim of the book is to show how the dominant geopolitical imagination arose from European–American experience but was then projected on to the rest of the world and into the future in the theory and practice of world politics. Beginning in the 1980s, as the Cold War ended, globalization began to offer the possibility of a world structured in a very different way. For example, as the Berlin Wall fell in November 1989, so also did the geopolitical template of Soviet East versus American West that had underwritten world politics for the previous forty years. National economic well-being increasingly seems to depend on advantageous connections to a world economy no longer totally dominated by the policies of dominant states. Even if the reactions by governments to any number of recent events – from the terror attacks of 11 September (and later ones such as that in Bali, Indonesia, on 12 October 2002) leading to a focus on Afghanistan and Iraq more than the terrorists themselves, to the continuing dispute between North Korea, on the one side, and Japan and the United States, on the other, and the bloody conflict between Russia and Chechen separatists as Russia reasserts itself as a Great Power defending its 'territorial integrity' – might suggest that the modern geopolitical imagination still has plenty of political mileage left in it yet, there are nevertheless trends whose interpretation points in a different direction.

More particularly, a number of crucial 'certainties' from as recently as fifteen years ago – about, for example, fixed and unquestioned political boundaries between states, the division of the world into mutually hostile armed camps on the basis of political ideology, the centrality of states to world politics, and the primacy of fixed national identities in political psychology – have either disappeared or are seriously in question. The end of the Cold War, the growing importance of trading blocs such as the European Union, and the proliferation of ethnic and regionalist movements within established states have served to undermine much of the conventional wisdom at the heart of the modern geopolitical imagination. Perhaps it is not entirely surprising, therefore, that there is now increased attention paid to the ways in which academics and political leaders have understood and practised world politics. In times of flux, conventional wisdom is more open to scrutiny.

This book focuses on the angle of vision provided to understanding world politics by the role of dominant geopolitical understandings. The phrase 'world politics' itself

Figure 1.1 A spectre haunts Central Europe, the history of its boundaries.

Source: Author

conveys a sense of a geographical scale beyond that of any particular state or a locality in which states and other actors come together to engage in a number of activities (diplomacy, military action, aid, fiscal and monetary activities, legal regulation, charitable acts, etc.) that are intended to influence others and extend the power (political, economic and moral) of the particular actors who engage in them. But the activities also rest on more specific geographical assumptions about where best to act and why this makes sense. The world is actively spatialized, divided up, labelled, sorted out into a hierarchy of places of greater or lesser 'importance' by political geographers, other academics and political leaders. This process provides the geographical framing within which political elites and mass publics act in the world in pursuit of their own identities and interests.

A couple of examples can serve to illustrate the ways in which the geographical framing at the heart of the modern geopolitical imagination takes place. Central Europe, for example, has actively figured in US, German, Soviet and other Great Power foreign policies for many years. It was the primary focus of Nazi German territorial expansion before and during the Second World War. Czechoslovakia was portrayed on Nazi geopolitical maps in the late 1920s and in the 1930s as a 'dagger' pointing at the heart of German territory. During the Cold War it was the main arena of 'geopolitical tension' between the US and Soviet 'sides'. Today it has re-entered world politics largely in relation to the geographical expansion of the European Union. The region's past roles, however, have not entirely faded. Indeed, the redrawing of international boundaries, loss of life, and large-scale population movements at the end of the Second World War still haunt Central Europe (Figure 1.1). In the summer of 2002, just as the European Union was discussing potential membership by various central European states, the rediscovery of the 'Beneš decrees' issued by the Czechoslovak government in 1945 (the legal basis for the eviction of the Sudeten Germans from Czechoslovakia) suddenly revived debate in Germany and throughout the region over what had become a 'dead' issue during the Cold War: the boundaries between the states of the region and the compensation (including restoration of property) due those displaced. Central Europe, therefore, has not entirely lost its role as a geopolitical marchland between West and East even as the historical context has obviously changed.

For a second example, in his State of the Union address in January 2002, President George W. Bush not only declared war on terrorism but also identified what he called an 'axis of evil' in world politics, with the three states of Iraq, Iran and North Korea as its geopolitical anchors (Figure 1.2). Formerly referred to, with a list of others, as 'rogue states', these three were singled out because of (i) their quest to possess 'weapons of mass destruction' – biological, chemical and nuclear; (ii) their hostility to the United States in particular and the contemporary global distribution of power in general; and (iii) their alleged support for terrorist groups and other rogue states. That Pakistan and Saudi Arabia, for example, might equally qualify for membership in the 'axis' because of the former's supply of bomb-making equipment to North Korea and its own pursuit of weapons of mass destruction and the latter because of its long-term support of fundamentalist Islam and financial support for bin Laden's *al-Qaeda* terrorists underscores the importance of criterion (ii) relative to the others. The image of an axis of states challenging the United States is more important than empirical accuracy in whether or not they are the most important developers of weapons of mass destruction or fomenters of terrorism. The image is that of an 'axis' or real connection between them, even though it is not clear what each shares in common beyond the three criteria listed above, and involves a clear analogy back to the Second World War when Germany, Italy and Japan (which really were allies) formed the so-called Axis Powers. President Bush was thus attempting to give a clear geopolitical structure to the threat faced by the United States (and the current world order) and recall a powerful threat from the past in support of his case. Later in 2002, President Bush invoked the 'axis' to justify a new foreign policy doctrine for the United States based on the possibility of conducting preventive wars against states that might eventually pose a threat to the United States (including the 'threat' to the world oil supply from countries such as Iraq in which the oil industry is state owned and not dominated by US companies) even if such a threat is currently not imminent.

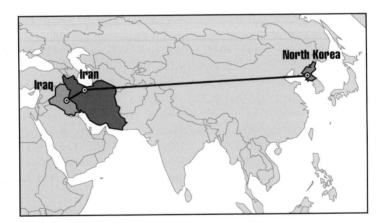

Figure 1.2 President George W. Bush's 'axis of evil'.

Source: Author

Indeed, one way of thinking about the geographical framing of foreign policies is to recall the foreign policy 'doctrines' enunciated by various US presidents down the years, from President Monroe in 1823 through President Truman in 1947 to President G. W. Bush in 2002. The Monroe doctrine initially involved three geopolitical imperatives relating to US foreign policy: the Americas were closed to further European colonization; the United States must avoid involvement in wars in Europe; and, in particular, that the US government would regard the effort by any European power to expand its empire into the western hemisphere as a threat to the United States itself. Eventually, for example during the Johnson and Reagan presidencies, this doctrine was invoked to justify US military intervention in various parts of Latin America – something remarkable by its absence from the original statement of the doctrine. In its turn, the Truman doctrine of 1947 helped to set the scene for the Cold War by challenging the Soviet attempt at violating the boundaries between the US and Soviet spheres of influence established just two years previously at the Yalta Conference by attempts at using surrogate forces to undermine governments allied with the United States, initially in Greece and Turkey but later more generally. The Bush doctrine of 2002, enunciated by President Bush in a speech at the US Military Academy at West Point, New York, is also explicitly interventionist, only the claim of the right to intervene now includes the explicit right to *preventive* US military intervention when a specific state is declared to threaten or pose an immediate danger to the US 'national interest' but has not yet mounted any kind of military strike.

The term 'geopolitics' has long been used to refer to the study of the geographical representations, rhetoric and practices that underpin world politics. The word has in fact undergone something of a revival in recent years. The term is now used freely to refer to such phenomena as international boundary disputes, the structure of global finance, and geographical patterns of election results. One expropriation of the term ascribes to it a more specific meaning: examination of the geographical assumptions, designations and understandings that enter into the making of world politics. Some recent work in political geography has tried to show the usefulness of adopting this definition of the term. In using this definition here, I attempt to survey how the discovery and incorporation of the world as a single unit and the development of the territorial state as a political ideal came together to create the context for modern world politics. Studying geopolitics in this sense, or geo-politics, so as to distinguish it from what is being described, means trying to understand how it came about that one state's prospects in relation to others' were seen in relation to global conditions that were viewed as setting limits and defining possibilities for a state's success in the global arena. The use of visual language is deliberate. World politics was invented only when it became possible to see the world (in the imagination) as a whole and pursue goals in relation to that geographical scale.

According to the history of the word, geopolitics began in 1899 when the word was first coined by the Swedish political scientist Rudolf Kjellén. The coining of the word meant that thinking globally on behalf of particular states was explicitly connected by formal geopolitical reasoning to their potential for acting globally. Ways of thinking and acting geographically implicit in the phrase 'world politics' had begun much earlier, however, when the intellectuals of statecraft of the European states (leaders, military strategists, political theorists) pursuing their 'interests' had to consider their strategies in terms of global conditions revealed to them by the European encounter with the rest of the world. Indeed, as Chapter 2 details, Renaissance-era Europeans drew upon older sources (particularly the cosmography of Ptolemy) to direct their understanding of the world. What was new, however, was that the world they encountered corresponded increasingly to the Earth as a whole rather than to the geographically limited worlds known to the ancients. The onset of the capitalist world economy and the growth of the European territorial state gave rise to a novel set of understandings about the partitioning of terrestrial space. The layering of global space from the world scale downwards created a hierarchy of geographical scales through which political–economic reality has been seen; in order of importance, the four have been the global (the scale of the world as a whole), the international (the scale of the relations between states), the domestic/national (the scale of individual states) and the regional (the scale of the parts of the state). Problems and policies have been defined in terms of the geographical scales (either domestic/national or international/foreign) at which they are seen as operating within a global context. Consequently, world politics is understood as working from the global scale down. It is at the global scale, therefore, that the term 'geopolitics' is usually applied. Yet it rests on assumptions about the relative importance of the various geographical scales to life on the planet that are already in place. The global and the national are privileged largely to the exclusion of the others. But naming such assumptions *geopolitics* and creating models that claim a superior grasp of geopolitical realities came much later than the onset of the geopolitical representations and practices themselves. The purpose of this book is to identify and critically examine the central features of the geopolitical imagination that arose in Europe, beginning in the fifteenth century, and that later became the stock in trade both of conceptions of world politics and of the practices of foreign policy.

The modern geopolitical imagination is a *system* of visualizing the world with deep historic roots in the European encounter with the world as a whole. It is a constructed view of the world, *not* a simple spontaneous vision that arises from simply looking out at the world with 'common sense'. As a system of thought and practice, the modern geopolitical imagination has not existed and does not exist in a material vacuum. It first developed in a Europe coming to terms with both a new global role and the disintegration of the religious-based image of universal order formerly dominant among its intellectuals and leaders. An insistence on taking charge of the world is a key feature of European modernity. Its realization has changed significantly over time as the material context (dominant technologies, modes of economic organization, scope of state organization, capacity for violence) has changed (see Agnew and Corbridge 1995). This is why after identifying and discussing its four foundations I turn to its historically specific operation over the course of the past two hundred years. This *historical* geopolitics, focused on practical geopolitical reasoning, needs to be distinguished from what is sometimes called *critical* geopolitics, taken either with exposing the hollow claims of particular geopolitical writers to having found the 'truth' in world politics or with identifying the representational basis to particular foreign policies.

Several other distinctions are also worth making, to clarify the purpose of the book. This is not a restatement or elaboration of the geopolitics associated with the inventors of the term and its applications: Rudolf Kjellén, Friedrich Ratzel, Halford Mackinder, etc. Neither is it an argument in support of that tired, cynical, aristocratic world-view popular with political realists in establishment universities and think tanks (for recent good examples of the genre, see, for example, Mearsheimer 2001; and Haslam 2002) that the general nastiness of the world means that there is 'no virtue like necessity' in constantly regarding other places (states and people) as threats and dangers. It is, rather, an analysis of what I regard as a historically constructed practical imposition on a complex and diverse world that is more accurately thought of as a mosaic of places linked and differentiated in multiple ways than as a simple reflection of the competing geopolitical 'interests' and 'identities' of fetishized states.

The modern geopolitical imagination at work: two examples

Two relatively familiar examples from recent world politics can help to show how certain geopolitical representations underwrite specific policies and interventions (practices) that are then interpreted in terms of these representations. The first example concerns the recent conflicts in the former Yugoslavia. The break-up of the former Yugoslavia along ethnic–regional lines in the 1989–94 period might appear to have had local causes relating to the distribution of power and economic well-being of the various parties in dispute (Serb groups, Croats, Bosnian Muslims, others). But outside actors, such as the US and German governments, framed their policies largely in terms of bigger geopolitical pictures. These policies were crucial to the course of the conflict because all the internal parties needed external allies. For the United States, maintaining the territorial integrity of the country was paramount throughout the early years of the conflict because Yugoslavia was seen as a buffer state between the Soviet Union and its allies to the east and the United States and its allies to the west. The then West German government adopted a very different position. Thinking in terms of the investment and trade potential of the wealthier regions of Slovenia and Croatia, the German government encouraged political independence for both these regions and Bosnia without requiring guarantees as to how minority populations (largely Serb) would be treated by the newly independent states. In both cases a geopolitical framing that reflected different globally defined interests on the part of powerful external actors contributed to the conflict, even though each was rhetorically devoted to resolving it.

Many commentators have made much of the geopolitics of the break-up of the country, paying little or no attention to the nesting of causes across geographical scales. This reflects the privileging of the global and national scales implicit in the modern geopolitical imagination. At the same time, much of the discussion about maintaining the original state or recognizing the various parts was conducted in terms of the absolute sovereignty of individual governments to do this or that within *their* territories. Statehood was the name of the game. The objectives of all sides were also expressed in terms of facilitating the entry of the area into 'the West', so as to lead it out of its economic and political backwardness. Finally, the various external parties (the United States, the European Union, Germany, France, etc.) were all concerned about the fall-out from their policies on their respective positions in the global pecking order of states. For example, the US government worried that the Soviet Union (as it was at the outset of the conflict) would gain political benefit from a divided Yugoslavia at US expense. Each and every one of these is a feature of the modern geopolitical imagination with a strong hold over political elites (and political geographers) around the world.

US military intervention in Vietnam in the 1960s was based on a similar set of geopolitical assumptions. The first was that US involvement was a necessary part of the geographical containment of world communism as represented by the Soviet Union and its allies, including the Viet Cong in South Vietnam and the Communist government in the north, as first enunciated in the years 1947–50. The second, and inseparable from the axioms of containment, such as domino effects – as one country after another would fall if a stand were not taken in the first place – and linkage – Vietnam could be understood only in relation to the overarching global conflict between the United States and the Soviet Union – and credibility – if a stand were not taken there, other (more important) allies would question US commitment to them – was that historical analogy pointed to the universality of these geographical 'truths'. US leaders used 'history' to justify their decisions, as if history simply repeated itself without the necessity of geographically informed judgement about the character of different world situations. The particular historical analogies were so familiar to the generation of US politicians who engineered American intervention in Vietnam that they could be communicated by single words: 'Munich' and 'Korea'. Each was shorthand for the future costs of appeasement of aggression or intimidation. Only if a Viet Cong mobilization were contained could a future bloodier conflict be deterred. Finally, 'North' Vietnam was portrayed as 'invading' 'South' Vietnam, so as to define the conflict as the result of an act of international aggression rather than a civil war between competing political factions who at one time (during the latter phases of French colonial rule) had operated in the country as a whole. But the partition of 1954, achieved largely at the behest of the US government, which refused to allow all-Vietnam elections after French withdrawal, became the crucial geographical representation in characterizing the war to the American population as a war of invasion that had to be resisted. In this account the Viet Cong was a 'fifth column' (a term first used in the Spanish Civil War of the late 1930s but going back to the Trojan Horse of Greek antiquity) acting as agents of the 'North' in the 'South'. Of course, as we now know, the 'loss' of South Vietnam following US withdrawal under heavy domestic political pressure did not lead to the Communist takeover of all South-East Asia. Key geographical assumptions proved fallacious, yet they were very much the synthetic elements in lending meaning to the intervention as a whole, as painfully related in the memoir of the then US Secretary of Defense, Robert S. McNamara (1995). More recent American military interventions in the 1980s and 1990s, such as those in Grenada, Kuwait and Haiti, to choose just three at random, though differing in detail from that in Vietnam, have been accompanied by the same kind of geographical rhetoric, including frequent reference to the 'lessons' of Vietnam (restrict media access, bomb endlessly, avoid slow escalation, etc.), indicating the extent to which US foreign policy on the whole became dependent upon a particular set of geopolitical assumptions during the Cold War period of 1947–89. These geographical assumptions and language indicate the facets of a more inclusive geopolitical imagination that insisted on dividing the world as a whole into units of sovereign statehood, the challenge and response of a particular superpower (the United States) to the pursuit of primacy (becoming 'number one') by other leading states (such as the Soviet Union), the characterization of South Vietnam as a 'backward' country that could 'develop' successfully only by imitating the American example of 'modernity', and the siting of Vietnam in a singular global context (that of the 'Cold War') rather than in terms of *local* causes of political resentment and revolution.

Global hierarchy and its origins

The fundamentals of the modern geopolitical imagination did not suddenly spring forth during the Cold War, whatever the specifics of the historical context. They had long-standing roots in the relationships established initially by European states among themselves and in their outlook towards the rest of the world. The first four chapters of this book are devoted to each of the fundamentals. An important caveat is that the geopolitical imagination as construed in the previous paragraphs has never exercised absolute power over the course of world politics, in the sense of transcending the effects of technological, economic and other material determinants, or in totally dominating understanding by political geographers and others. The modern geopolitical imagination has, however, provided meaning and rationalization to practice by political élites the world over. It has defined the 'ideological space', to use Immanuel Wallerstein's (1991) phrase, from which the geographic categories upon which the world is organized and works are derived. From this point of view, the history of modern world politics has been structured by practices based on a set of understandings about 'the way the world works' that together constitute the elements of the modern geopolitical imagination. It is Eurocentric because Europe and, later, its offshoots (such as Russia and the United States) came to dominate the world. Political élites around the world adjusted to and adapted understandings and practices emanating from Europe. A large part of Europe's heritage to the world is the continuing strength of the 'common sense' about world politics bequeathed in the form of the modern geopolitical imagination. From Brasilia to Seoul and from Cairo to Beijing the dominant model is still that invented originally in Europe.

This does not mean that there have not been other causes of world politics to which the geopolitical imagination has had to relate and adjust, such as technological and economic changes. For one thing, the rise of European and US power has had its material roots based on, at minimum, economic and social pressures to expand geographically, military and naval technological advantages, diseases to which other people had limited or no immunity, and an ability to borrow from and synthesize elements of other cultures. For example, as the industrialized world has become increasingly dependent on oil, the map of world oil supplies has become an increasingly important part of the calculus driving the operation of the geopolitical imagination. At the same time, however, the importance of the rise of the modern geopolitical imagination does not mean that oppositional ideological currents (such as anti-colonialism, laissez-faire economics and some varieties of internationalist socialism, for example) have failed to challenge the dominant representations and practices; only that these have had to contend with a powerful and well-established conventional wisdom about 'how things are' that has reproduced the state-based and place-hierarchy view of the world that lies at its heart. Movements to remove colonial rule often ended up accepting and sometimes celebrating the often problematic political boundaries imposed by colonial empires. India and the countries of Africa are classic cases. 'Socialism in one country' (as in the former Soviet Union or in contemporary China) and neo-liberal economic policies based on activist governments (as with the Reagan administration in the United States, 1981–88, and the Thatcher government, 1979–90, in Britain) reflect the channelling of oppositional discourses within parameters defined in large part by the hegemonic geopolitical imagination.

The dominant representations and practices (or *hegemony*) constituting the modern geopolitical imagination have been overwhelmingly those of the political élites of the Great Powers, those states and empires most capable of imposing themselves and their views on the rest of the world. Membership of this group depends on recognition by existing members. Qualification has depended not simply on coercive power, the ability to force others to do what you want, but also on the capacity to write the political–economic agenda of others, defining appropriate standards of conduct and providing the framing for inter-state relations with which others must conform if they are to gain recognition and rewards from the Great Powers. News stories using the 'common-sense' accounts of what is at stake for 'us' or others in this or that part of the world, 'official stories' told by political leaders, and the presentations of intellectuals elaborating on the 'logic' of particular foreign policies and military strategies are all important ways in which the dominant story-lines and agendas of different states (and other actors) can be disseminated, both domestically and internationally.

At any one time a single state can be number one among the Great Powers, having a key role in laying down and enforcing the 'rules of the game' (which can be referred to by the term 'hegemony'). This was the case with Britain in the mid-nineteenth century and with the United States from 1945 to the 1970s. With the Soviet collapse in the early 1990s the United States has acquired an even more obvious hegemonic position. But as long as political élites engage in relations with one another there must be rules governing their interaction. So, even in the absence of a particular hegemon there is still hegemony in the sense of a set of dominant understandings and practices regulating world politics. Recently, a transnational liberalism sponsored by governing élites around the world, but with particularly important American sponsorship, has seemed to offer an emerging ideological hegemony without the necessity of a single hegemon (or dominant state). Based more on principles of free trade and comparative advantage working to the benefit of global business enterprises (and their local allies) than on the enhancement of state power, its putative existence might call into question many of the established tenets of the modern geopolitical imagination. Its emergence does not signal the 'end of geopolitics', however, so much as its possible reformulation from a state- towards a firm- and city-centric focus in which (some) states must adjust to new demands on them to relax their controls over markets. As yet, however, the modern geopolitical imagination, though changing its emphases as the balance of power among the Great Powers and the nature of the world economy have changed over time, still remains prevalent in framing the conduct of world politics. Not surprisingly, political geography and adjacent fields of study such as international relations and political sociology remain largely attached to the same vision. What are the roots of this imagination? Why has it had such staying power? These are among the questions addressed in this book.

Outline of chapters

The first four chapters focus on key foundations of the modern geopolitical imagination and how it has developed down the years. The first chapter addresses a primary feature: a global vision without which *world* politics would be impossible. The acquisition and perpetuation of this vision – the sense of a world-as-a-whole that powerful actors must survey and subdue – are traced through the history of early modern cartography, the theorizing of 'one world of humanity', imperial cosmologies, colonial economics, formal geopolitical models and the global ideological polarization of the Cold War. The purpose is to show the cumulative historical basis to global geographical visualization: how the Earth was made into the World. Thinking about the world as a whole was not a one-time thing, established once and then taken for granted. It had to be constantly reproduced in changing economic and technological circumstances to remain effective. Seeing the world as one and then dividing it into a hierarchy of places have required relating thinking about the world to changing material conditions even as the a-priori assumption of wholeness is perpetuated. Encapsulating the world in the mind's eye as the framework for understanding anything that happens anywhere also can serve to legitimize what could be a very partial view of the world, from one's own position at a particular location, into a view that could claim to be *a view from nowhere*: an objective view of the world as from outer space. This is not to say that such a view is a-priori illegitimate intellectually and politically, only that the association of global geopolitical thinking with the self-defined interests of specific states (such as Germany or Britain) tended to use the scientific claim to objectivity on behalf of a particular identity/interest. This use of the view from nowhere has always been an important part of the modern geopolitical imagination. Its relating of anywhere to everywhere rests on the legitimacy derived from its being a view from nowhere. If it were seen as just a view from somewhere, its partiality and situatedness would be signalled from the outset. In this way a political claim on behalf of a particular interest is turned into a natural claim about the world as it is.

A fundamental feature of the contrasts drawn between different parts of the world seen as a whole has been the labelling of blocks of global space as exhibiting the essential attributes of the previous historical experience of the dominant block. This translation of 'time into space' is the subject of the second chapter. Typically, modern geographical taxonomy involves the naming of different world-regions or areas as 'advanced' or 'primitive', 'modern' or 'backward'. Europe and some of its political–cultural offspring (such as the United States) are seen as defining modernity, and other parts of the world figure only in terms of how they appear relative to Europe's past. Being like or imitating Europe thus becomes a condition for entry into the state system (as opposed to justifying subjugation) and provides the norm or standard for judgement about particular states (who is most advanced?, etc.). The case of the idea of Three Worlds of Development will be used to show how the global partitioning of space according to temporal stages of development has become woven into world politics. The exploration and early settlement of Australia are used to illustrate the ways in which an existing understanding of that continent's place in the world informed the outlook and behaviour of the explorers and settlers. The case of the historiography of Italy, a European state often seen as lagging behind the rest of the continent, illustrates how pervasive the use of time-into-space judgement has been in the making of the modern geopolitical imagination.

For world politics the fundamental world map is a world political one: the map of territorial states. The third chapter identifies the geographical assumptions that produce this state-centric aspect of the geopolitical imagination. Three assumptions are seen as integral to the state-centricity of the modern geopolitical imagination: first, state sovereignty and territorial space; second, the territorial state as container of society; and finally, the domestic/foreign polarity. These assumptions are modern ones; the first dates from the sixteenth century at the earliest and the others are even more recent. Yet the idea of a world made up entirely of states or territorial actors, as opposed to other forms of polity, is given a transcendental force within the geopolitical imagination. In world politics space can only be divided between actual states and states-in-the-making. Other modes of organizing politics geographically (such as kin-based polities, city-states, classic empires such as those of Ottoman Turkey or Manchu China, and confederations) are seen as either relics or 'really' territorial states in disguise (they can be regarded 'as if' they were real states). Territorial states are the individual actors of the geopolitical imagination. This perspective, cleaning up the large variety of polities to be found in all epochs into a single type, is often referred to as the 'Westphalian view', after the Treaty of Westphalia (1648), which settled the religious wars between Catholics and Protestants in Europe by establishing as a standard the idea of a single governmental jurisdiction over a single territory.

The dynamic force bringing together the other elements of the modern geopolitical imagination is devotion to the idea that pursuit of one state's (typically, mine or your) interests or *security* relative to those of all others is necessary for personal ontological security. The modern world is seen as one of unremitting competition for primacy: either to dominate the world economically (in some recent understandings) or to turn it into a world empire (in more typical usage). The fourth chapter addresses three questions that arise from this conception of world politics. First, what makes a Great Power a potential hegemonic state? This directs attention to two crucial features of thinking about interstate competition: uneven rates of economic growth and international anarchy. The two other questions follow: under what historical–geographical conditions has the pursuit of primacy been possible? And is it still possible at the outset of the twenty-first century?

Identifying these 'fundamentals' is all very well. But how have the geopolitical underpinnings of modern world politics changed over time? Chapter 6 provides a narrative account of three 'Ages of Geopolitics' in which the modern geopolitical imagination has shown distinctive features and relations to practice. The first, prevalent in the eighteenth and early nineteenth centuries, was a *civilizational geopolitics*, in which Europe's unique civilization compared to the newly discovered rest of the world played a central role. The second, dominant from the late nineteenth century to 1945, was a *naturalized geopolitics*, in which the 'natural' character of states as predators and competitors assumed a key position. The third, operational during the years of the Cold War, an *ideological geopolitics*, was based on dividing up the world between competing ideas about how best to organize political and economic life ('socialism' versus 'capitalism', etc.). Although the principles of geopolitics as laid out in Chapters 1–4 remained in effect throughout, the net impact was not always the same. Within a general continuity, therefore, one can identify distinctive epochs in which the geographical representations and practices implicit in world politics have undergone important shifts.

Finally, Chapter 7 considers what recent great political changes, such as the end of the Cold War, technological and economic changes, such as those associated with an emerging informational capitalism and globalization, so-called civilizational conflicts, and the singular superpower role of the United States, might augur for the future of the geopolitical imagination and for the political geography that remains attached to it. Is a new epoch of postmodern geopolitics on the horizon or, as I think more likely, an emerging geopolitical order dominated by a US attempt at pushing globalization but frustrated by US national economic weakness, widespread nostalgia for a putative past of 'secure' national territories (not least in the United States), and an inability to come to terms with non-state adversaries such as global terror networks and drug traffickers and to deal easily with disintegrating state sovereignties and the humanitarian crises that follow? Given how much political élites and many intellectuals of all political stripes remain attached to it, the modern geopolitical imagination seems set to create more woe than progress for the world in the years ahead.

The concluding chapter (Chapter 8) identifies the crucial premises of the approach taken in this book and provides a review of some of the criticisms that might be addressed to it. I then draw together the arguments of previous chapters to claim that understanding global political geography stands in need of recomposition beyond the frame of reference supplied by the modern geopolitical imagination. This new approach would resist privileging certain geographical scales of analysis, such as the global and the national, as supposedly of singular importance. Rather, rethinking geopolitical analysis would lead to an emphasis on how different scales and networks between places relate to one another differentially over time. This moves thinking away from state-centred political horizons towards a more pluralistic vision of political organization in the past and in the future. This alternative imagination might help in thinking about and starting to practise a world politics no longer subject to the depredations of the modern geopolitical imagination.

By way of summary for the book as a whole, world politics can be analysed in terms of a set of principles that collectively constitute the modern geopolitical imagination. But these must also be put together in terms of how they have operated in different historical epochs. By using an analysis of the modern geopolitical imagination as a way of interrogating world politics, therefore, a new narrative structure can be given to modern world history (Chapters 6 and 7), in the process also identifying the critical geopolitical understandings and practices upon which world politics and many approaches to understanding it continue to rely (Chapters 2–5). That it is no longer suffi-cient, even if one grants it a certain logic or at least inevitability in previous epochs, and now needs to be replaced by a different approach to understanding the world's polit-ical geography and thereby to begin making a new world politics is the implicit theme of the book as a whole.

Recommended reading

Agnew, J. and Corbridge, S. 1995 *Mastering Space: Hegemony, Territory and International Polit-ical Economy*. London: Routledge.

Alker, H. 1996 *Rediscoveries and Reformulations: Humanistic Methodologies for International Studies*. Cambridge: Cambridge University Press, chapter 5.

Campbell, D. 1997 *National Deconstruction: Violence, Identity and Justice in Bosnia*. Minneapolis: University of Minnesota Press.

Connolly, W. 1989 *Political Theory and Modernity*. Oxford: Blackwell.

Haslam, J. 2002 *No Virtue Like Necessity: Realist Thought in International Relations since Machiavelli*. New Haven, CT: Yale University Press.

Heffernan, M. 1998 *The Meaning of Europe: Geography and Geopolitics*. London: Arnold.

Klare, M. 2002 Hawks with their eyes on oil. United States: energy and strategy. *Le Monde Diplomatique* (English-language edition), November: 1–2.

McNamara, R. J. 1995 *In Retrospect: The Tragedy and Lessons of Vietnam*. New York: Times Books.

Mearsheimer, J. J. 2001 *The Tragedy of Great Power Politics*. New York: Norton.

Ó Tuathail, G. 1996 *Critical Geopolitics: The Politics of Writing Global Space*. Minneapolis: University of Minnesota Press.

Record, J. 2002 *Making War, Thinking History: Munich, Vietnam, and Presidential Uses of Force from Korea to Kosovo*. Washington, DC: Naval Institute Press.

Smith, G. 1994 *The Last Years of the Monroe Doctrine, 1945–1993*. New York: Hill and Wang.

Toulmin, S. 1990 *Cosmopolis: The Hidden Agenda of Modernity*. Chicago: University of Chicago Press.

Wallerstein, I. 1991 *Unthinking Social Science: The Limits of Nineteenth-Century Paradigms*. Cambridge: Polity Press.

2 Visualizing global space

An account of the modern geopolitical imagination – the predominant ways world politics has been represented, talked about and acted on geographically by both major actors and commentators over the past two hundred years – must start with the origins and development of the capacity to see the world as a whole. From this point of view, the 'modern' world is defined by the imaginative ability to transcend the spatial limits imposed by everyday life and contemplate the world conceived and grasped as a picture. The geopolitical imagination, therefore, is a defining element of modernity. Its most distinguishing feature is the conception of the world as a single if divided physical–political entity – a feat of imagination impossible before the encounters of Europeans with the rest of the world beginning in the late fifteenth and early sixteenth centuries.

Two characteristics of visualizing global space emerged at the outset of the European Age of Discovery. Ever since, they have been reproduced in the governing principles of geographic thought and through the practices of statecraft. The first, and the one that has received most recent comment, is that seeing the world-as-a-picture, as an ordered, structured whole, separates the self who is viewing from the world itself. The observer stands outside of terrestrial space, so to speak, and frames the world as apart from and prior to the places and people it contains. This seems to be a peculiarly European perspective in origin, associated with the Renaissance-era separation of the observer from the world and an approach to knowing that insists on privileging vision as the 'noblest' of the human senses. What is 'seen', even from over the horizon by means of such tools as world maps, is what exists. The map is an accurate report of what is there. Representation and world are as one. In reaction to this ennobling of singular vision and the absence of scepticism about its impact, much recent philosophy has become deeply suspicious of vision and what Martin Jay (1993, 14) terms 'its hegemonic role in the modern era'. This doubt can be traced to the philosopher Friedrich Nietzsche's (1844–1900) and more recent feminist and other critiques of the 'view from nowhere'; there can be no pure, will-less, timeless, placeless knower, no pure reason, no absolute knowledge or absolute intelligence. All seeing and knowing is a perspective, drawn from a situated point of view. From this contrarian standpoint, real objectivity does not lie in the promulgation of a single perspective from within a singular historical experience but arises from the introduction of as many eyes (perspectives) as possible. There is never a single, objective view from nowhere.

The second characteristic of global visualization is that the world pictured beyond the horizon is a source of chaos and danger. The evil spirits and dark places decorating the borders of early modern maps, where they signified 'unknown' and presumably dangerous places, moved into the world itself where they came to represent fearful religious, civilizational and political differences. The one world-picture, therefore, is not a composition of equal and pacific elements but a hierarchy of places, from known to unknown, from most friendly to most dangerous. The best-known representation of this character is that of a dichotomous global West and East, in which the former is seen as the total opposite and, hence, definitive standard for the latter. With roots in an ancient European past, this opposition serves as a geographical template on to which more local differences can be mapped. They then become explicable only as elements in the bigger picture. Local differences are ascribed to worldwide distinctions rather than to local differences per se. The local has meaning *only* in relation to the global. Without the global reference, difference cannot be articulated. This is the attribute of linkage whereby specific places are drawn into an overarching global geopolitical frame of reference.

The idea of the West as a civilization rests on this foundation. It distinguishes Europe and its overseas offshoots, particularly the United States, as fundamentally different from the rest of the world. In fact, defining and defending 'the West' in these terms is a civilizational project – an attempt at providing a gloss of cultural homogeneity for an otherwise geographically divided and historically conflict-ridden part of the world. Today the term is often applied entirely to the actions of US governments, which are presented as speaking for the West as a whole. But its origins lie in the binary geography initially identified as Europeans encountered a world previously beyond their experience.

This chapter is about the origins and elaboration of the twin features of global visualization. I begin with the Renaissance origins of the view of the world from beyond its limits: the view from nowhere. A case is then made for how this perspective (in both theoretical and visual senses of the word) was both elaborated on and familiarized by the empiricist conception of knowledge and its commitment to vision as the primary human sense. With respect to the second feature, I identify some of the sources of binary geographical thinking, particularly the East–West opposition, and argue for how these have been reinforced historically by means, for example, of imperial cosmologies, colonial economics, formal geopolitical schemas, historical analogies, and informational technologies. Coming to see the world both as a whole and as dangerous was not a one-time thing, neatly packaged at one moment and thereafter reproduced in exactly the same way without ambiguity or challenge. Visualizing global space has involved the adaptation of its two features within the overall context of a persisting modern geopolitical imagination.

Seeing the world as a structured whole

Events more than ideas capture immediate popular attention. So it was neither Galileo's refutation of geocentrism – the assumption that a stationary, immovable Earth was the centre of the cosmos – nor the anonymous discovery that the Sun is a star that produced the realization that the Earth is a rotating spherical object on whose surface everywhere is connected to everywhere else. It was perhaps the return of Magellan's crew in 1522 from the first known circumnavigation of the Earth that really brought this home. The Portuguese explorer never made it back himself; he was killed in what later were named the Philippines (after King Philip II of Spain). 'The Europeans', to quote the historian J. H. Elliot (1991, 10), 'had found space, and found it on an unimagined scale. But, paradoxically, even as their world expanded, it also began to shrink. A globe encompassed became a globe reduced.' Thereafter, Europe was no longer the world and the world was no longer the centre of the universe. This had revolutionary consequences for both the European world-view and the meaning of life:

> Since the earth was revolving daily, heaven and hell could not be located where they had been thought to be, and in rational minds there was a growing skepticism that either of them existed. Satan without hell was implausible. God without heaven was inconceivable, at least the medieval God was, but here reason ended.
>
> (Manchester 1992, 295)

The familiar vertical conception of the universe (the Great Chain of Being) that connected ordinary mortal Europeans into the universal scheme of things was challenged by an opening up of horizontal horizons that offered a new vision of the world and Europe's place in it. An alternative framework for understanding the new world was needed and, in basic outline, it was fast in coming.

The Renaissance rediscovery of Ptolemy (who lived AD 90–168) provided a suitable model of the structure of the world. Summarized, used as a model and popularized above all by Sebastian Münster in his *Cosmographia universalis* (1550), Ptolemy's cosmography (an imagined world structure into which phenomena for study in their own right could be slotted) supposed a global world with no limits other than those of its poles, regions and zones (Figure 2.1). This model was inadequate in many of its details, as would be revealed over the next several hundred years. For example, in its account the glacial and torrid zones were uninhabitable by humans, there was exact symmetry between the northern and southern hemispheres, and America had no place; stretching as it did from north to south and in two blocks joined by a land bridge, it was also organized in a very different way from that of the Old World's division into the three continents of Europe, Asia and Africa. But Ptolemy's model offered a ready substitute for the dominant medieval world-view. Its main advantage was its openness to the filling in of the 'unknown' spaces as more voyages led to the enlargement of perspective and the impetus for trade and conquest prompted speculation and inquiry about 'unvisited' parts. Most of all,

> this model, apparently so inadequate, proved to be fecund by virtue of its own anachronism. It offered to modern geographers a three-quarters empty canvas, leaving them free to inscribe on it the delineation of newly 'invented' or discovered lands; a form, at once closed and open, full and lacunary, that represented the ideal construction in which to house, with their approximate and disparate localizations, the 'bits' of space that navigators brought back from their distant voyages.
>
> (Lestringant 1994, 7)

On this open 'canvas' could be played out the 'cosmographical fictions', such as the search for the North-West Passage or a southern continent, that inspired so many of the personal and national projects of exploration and conquest. Many of these 'fictions' turned out to be hoaxes, based on claims without any foundation of one sort or another, but others turned out to be serendipitous, encouraging further exploration and discovery irrespective of their erroneous or lunatic origins. A good example of a probable hoax that proved serendipitous is Sebastian Cabot's 'expedition' of 1508 that presumably initiated the search for the North-West Passage around North America as a route to Asia. A recent study offers compelling evidence that 'Cabot's whole voyage was fictitious, that in fact he never left England' (Roberts, 2001, 6).

The expansion of speculation and knowledge about the world on a Ptolemaic basis is most clearly recorded in the world maps of the sixteenth to nineteenth centuries. In the early sixteenth century Ptolemy's original maps were generally predominant in Europe. Very quickly, however, national rivalries underwrote the efforts of cartographers to keep up with (and sometimes to anticipate!) the process of discovery. As early as 1489, Henricus Martellus, though still following Ptolemy, introduced a new coastline for South Africa to take into account the voyage of Diaz three years previously. By 1506 Giovanni Contarini showed a complete outline for Africa and noted the discoveries of Columbus. In 1507 Martin Waldseemüller 'envisioned an absolutely insular America more than 270 years before Cook finally demonstrated that it is indeed fully detached from Asia' (Zerubavel 1992, 84). In 1529 Diego Ribeiro made use of Antonio

Figure 2.1 Ptolemy's world outline, from the Rome edition of 1409. Originally compiled at Alexandria, Ptolemy's guide to geography was preserved for posterity by Arab and Byzantine scholars. Translated into Latin from a Greek manuscript in Florence in 1409, printed editions circulated widely in Europe only in the last decades of the fifteenth century.

Source: Crone (1978)

Pigafetta's report of Magellan's journey to construct a new world map. Finally, Blaeu's map, published in the Netherlands in 1648 to celebrate the Peace of Westphalia and symbolically representative of the state-centred world then in ascendancy, showed much of the outline of Australia and New Zealand as reported by Abel Tasman in his voyage of 1642–43. Thus, by the middle of the seventeenth century European scholars and political leaders already knew that they lived in a world that had a definite shape, even if actual knowledge of all of it was still incomplete by twenty-first-century standards.

These maps, speculative as they often were, share two features. One, common also to small-scale maps in previous and later epochs, was a pervasive ethnocentrism. The maps are centred on Europe (notably in the long-popular Mercator map), conveying a clear sense of Europe's geopolitical centrality. There is nothing passive about a map or a compilation of maps in an atlas. Maps can convey a strong message on behalf of a particular world-view. For example, eighteenth-century British maps of the American colonies included insets and cartouches that vilified the French, stereotyped African slaves, removed Indians cartographically before they could be removed physically, and showed the extent of British dominion over an inviting landscape. The second feature is their hierarchical representation of space, identifying and naming sites in terms of social and geopolitical significance. The power of the maps lies in their masking of these features behind a veneer of objectivity. The selectivity that went into the construction of the maps is never made evident. The reader has to presume a fairly close approximation between the maps and what lay 'out there' in the world beyond immediate experience. The maps convey an image of a world without the actively mediating hand of the cartographer. What appeared was surely what was there?

The modern cosmographer had a number of differences from the ancient one, therefore. The first was with respect to the perception of space. Perception of space was no longer totally abstract. The horizon could now be crossed. This gave a sense of both ubiquity and omnipotence that Ptolemy had lacked. The world now could be both thought of and experienced as a whole. Indeed, the practice of navigation over the open sea required a precise positioning of the travelling self in relation to the world as a whole. The second was the increasingly direct experience of the world, which gave travellers the authority to speak about where they had been in terms of their determination of its place in the world. Having been somewhere now provided a licence to speculate about everywhere. The ancients could only conjecture; the new knowledge of the world was based on an amalgam of personal vision, fantasy and speculation. So while drawing on an ancient Greek source for its model, modern cosmography combined it with both the astrolobe and the globe of the navigator and the naïve experience of the observer–writer. The third, reflecting the etymology of the Greek word *kosmos*, the root of such words as cosmography and cosmetic, was the aesthetic attention given to the singularity or beauty of objects. At the start, as in the case of the French cosmographer Thevet, it was only the uniqueness of what was discovered in terms of its absolute empirical variety that mattered. The wonder and novelty of the world out there and the sheer amount of new information sometimes resisted open theoretical classification (Plate 2.1). But with the rise of specific intellectual disciplines devoted to this or that phenomenon, order was imposed on the immense variety of the world by drawing on what was thought of as the experience and example of Europe itself. The form that this imposition of order took is examined in Chapter 3.

Plate 2.1 A German woodcut of *c.* 1505 showing a representation of the Topinambas of coastal Brazil. The picture combines a confusing mixture of routine cannibalism and domestic life which suggests that its creator was not quite sure what to make of stories about the Topinambas and erred on the side of including everything he had heard.

Source: The Spencer Collection, New York Public Library, Astor, Lenox and Tilden Foundation

The cosmographical model, the projection of a presumed structure to the world's physical geography drawn from ancient sources, was not to last in unrefined form. Very quickly, more specialized studies of geography (large-scale global patterns of seas, continents and climates) and chorography (local regions and landscapes) replaced the focus on the Earth within a larger universe implied by cosmography. But it had provided an orientation to the world as a whole upon which later generations were to build. Travellers carried with them its view of 'Spaceship Earth' even as their adventures slowly undermined many of its geographical presuppositions. In the first place, it replaced the Creator with the 'seeing' observer who in the name of experience spoke from authority. It established the view from nowhere as the modern ideal. It also used partial knowledge (what today we might call 'fieldwork') to justify knowledge of the whole. Having been somewhere was equivalent to having been everywhere. Images rather than words, or what geographer Denis Cosgrove (2001, 28) calls 'a distanced, rational gaze', gave authority to both the geography and the chorography that descended from the older cosmography. Finally, the cosmographical model saw the world as a whole – a canvas upon which a modern, Europe-centred world could be painted. Finding out was not so much about understanding in local terms as about compiling and collecting so as to incorporate the world's variety into categories familiar to the European mind.

The European Renaissance, however, had another influence beyond that of the revival and elaboration of Ptolemy's cosmography. This was the rediscovery of 'perspective'. Brunelleschi, the great architect of the Duomo in Florence, is technically the author of that optic of artistic representation in which a structure or a scene is viewed in a linear plane which the architect or artist uses to communicate to the observer an intended meaning. Thus a consistent reading of a building or a painting is made possible by imposing an orderly vision by means of an external point of viewing from which an object always appears in the same light. This conception of the consistent gaze as the key to understanding was not restricted to the pictorial arts. The linear perspective was an integral part of the Renaissance world-view. The language of knowing – perspective, view, vision, world-view – was expressed in overwhelmingly visual terms. The world was an object separated from the viewing subject, and could be understood adequately only if seen as a whole. Consequently, 'parts' are understood only in relation to the whole. Perspective allows a framing or 'field of projection' of particulars as elements in an ordered whole. John Berger (1972, 16) has summarized the implications of the rise of perspective as the governing principle for both viewing and knowing:

> The convention of the perspective, which is unique to European art and which was first established in the early Renaissance, centres everything on the eye of the beholder. It is like a beam from a lighthouse – only instead of light travelling outward, appearances travel in. The convention called those appearances reality. Perspective makes the single eye the centre of the visible world. Everything converges on the eye as the vanishing point of infinity. The visible world is arranged for the spectator as the universe was once thought to be arranged by God.

The idea of a uniform space falling away from the observer underwrote this vision. Within it a visual field would have definite limits, but what is seen would always be conceptualized with reference to it. This fitted into the evolving economy of the times. Objects such as those in a visual field had no intrinsic value outside of their relations to that field – equivalent to the exchange value of goods in capitalist trade. The rise of perspective, in liberating knowing from obedience to past texts, therefore, also brought with it an orientation towards objectification of the world in the interest of trade and commerce. This allowed the world to be divided up into a variety of grids and reproduced using a number of projections. This quantification of the world into precise bits and pieces (by latitude and longitude, for example) was possible only because of the visualization of the world as whole. Powers of 'planetary scope' could henceforth be deployed to control sea lanes and routes of circulation. In this way, global space was 'hierarchized' and ordered into zones of greater and lesser significance in which different activities and behaviours were appropriate. Technical systems for ordering the Earth's surface, therefore, were important elements in the development of the modern geopolitical imagination.

Because it is now so familiar, the perspectival conception of knowing seems unremarkable. Yet it involved a revolutionary shift in consciousness about the relationship of self to the world (and others), even if it had earlier roots. The self was now external (as observer) but also in the world (as actor). To operate as the latter required the sense of the former. Previously, this separation between knowing and doing did not exist. Rather than being based on the particulars of local knowledge, modern knowing required a prior framework into which knowledge about people and places could be pigeon-holed. Order could be discovered through creating distance between observer and observed and then seeing what is observed in relation to a whole setting. The European observer, trading and conquering as well as looking, saw the world 'as a differentiated, integrated, hierarchically ordered *whole*' (Gregory 1994, 36), in which he (it was usually a he) was both the external arbiter of truth about the nature of the parts and the primary agent in creating an integrated world where none had existed before.

The standard scientific model of knowing was built on this perspectival basis. Scepticism about the understanding to be gained from an explication of texts or great books (perhaps emanating, ironically given the scriptural claims on which it was based, from the Protestant Reformation's questioning of traditional authority) led to an emphasis on the virtue of 'direct observation' of facts that could be mapped and then subject to independent verification. An ethic of cumulative achievement of accuracy removed the inevitable selectivity of all maps or other representations of the world (however sophisticated) from active consideration. Rather than simply mirroring the world, therefore, popular maps helped to constitute it. In masking their selectivity behind empiricist claims to accurate representation they provided a powerful means of picturing the world as a whole as if it existed independently of or separately from all attempts to conquer, tame and exploit it.

There was no inevitability about drawing the connection between wholeness and hierarchy, however. Seeing the world whole, however selective in practice, could imply a sense of common fate, of a common humanity drawn together in planetary harmony, whatever local differences there might be. This was indeed the 'vision' of some Europeans. The Enlightenment cosmopolitanism of the German philosopher Herder (1744–1803) represents, for example, a version in which there is only one human race and a single human reason; climate and other geographical features are taken to account for such differences as exist between cultures. In this understanding, there can be no innately superior cultures and no groups or nations whose members are more fully 'human' than others. But European experience still provided the standard for what 'one humanity' might entail. Indeed, European experience defined what 'humanity' could be: a singular human nature resting on 'universal' attributes that were more often than not projections of European ones.

In the end, more hierarchical positions prevailed, ones more in tune with the growth of capitalism and the development of the territorial state. Increasingly, uneven development following the advent of a world market and European industrialization, the pairing of modernity with backwardness in understanding how this occurred (see Chapter 3), and the definite hierarchy among states (see Chapters 4 and 5) acted to reinforce the sense that places were hierarchically ordered on a world scale. In particular, the nineteenth-century Romantic reaction to Enlightenment rationalism added to the established perspectivalism notions of 'national character' which claimed superiority in human status and viewing capability for particular European nationalities. As Britain became the world-hegemonic power, English 'common sense', based on empiricist reverence for the accuracy of representations (such as maps) in directly revealing reality beyond the horizon, was established as the standard for judging 'things as they are'. Technological innovations such as the steamship, the telegraph and the telephone, and associated techniques for ordering the world into time zones (to make possible train and steamship schedules), reinforced the sense of a 'closing world' that could be represented and understood in a straightforwardly empirical manner.

Knowing by seeing the world as a *horizontal* but hierarchically organized whole was finally institutionalized as the modern alternative to the *vertical* Great Chain of Being (connecting the supernatural to the human worlds) that had been the dominant cosmology of other, older civilizations. The particular content might differ as intellectuals of statecraft in different states fused their own national myths into the general fabric of the geopolitical imagination. But the tendency to consider strategies and options within a whole-world framework was well established by the early nineteenth century across all the Great Powers and also those states that aspired to such status.

Binary geographies

It is well known that all societies define boundaries between themselves and others. The 'world' beyond the horizon is threatening as well as alluring. The zone of difference and danger is both repelling and attracting. At the same time, societies can exist only by defining themselves against an external standard – an Other without which the Self could not see itself as distinctive. 'European' society, the idea of Europe itself as a coherent social–geographical entity, could arise only in reference to what it was not and in relation to where it started and ended. Within Europe, and likewise, the nature of different national societies could be known only with respect to others they were not. Not all societies, however, are capable of imposing their maps of difference and danger on others. This is where material resources come into play. Europeans have been able on a world scale to do what only great empires, such as the Roman and Chinese, did more parochially: impose geographical boundaries on everyone else for a long period of time. In this way, local differences have been invariably assimilated into a global geographical taxonomy with its roots in Europe.

The most basic division was between Europe and the other continents. The oldest surviving European map of the world (from the seventh century AD) identifies three continents, Asia, Europe and Africa, does not have a surrounding ocean (as older ones probably did) and shows a large uninhabited area to the east. The boundary between Africa and Europe is the Mediterranean, that between Africa and Asia the River Nile, and that between Europe and Asia the River Tanais (the River Don) and the Meotides Paludes (the Sea of Azov). The map is, in the literal sense, 'oriented', aligned to the east, where the sun rises and where Paradise was supposedly located. The idea that the sons of Noah populated the three continents after the Great Flood is reflected in the naming of each on the map. Shem receives Asia, Ham Africa, and Japheth Europe. Thus, the Jewish and Christian story from Genesis of what happened after the Flood is incorporated into the map with Japheth (the precursor of Christian Europe, according to the dominant commentary of the time) identified with the Christian continent. This theme was muted, however, until the fifteenth century, when the identification of Europe with Christendom became commonplace. At this time the boundary between Europe and Asia moved considerably to the west. This was the result of divisions within Christianity (the Orthodox–Roman split), the resurgence of papal authority in western Europe following the Great Schism of 1378–1418, and the threat posed by the expansion of the Ottoman Turkish Empire. These conspired to produce a civilizational definition of Europe in which the term *Respublica Christiana* was applied to Europe and the adjective 'European' was used for the first time (both by Pope Pius II, 1458–64).

This sense of cultural unity was reinforced by two related intellectual trends. One was the spread of an ideal of education based on reading Greek and Latin (so-called Classical) authors in the original languages. This humanistic education linked the élites in European countries at the same time that they were becoming politically divided by the emerging merchant capitalism and state formation of the time. The second was the tendency to reach back to the Greeks and Romans to give Europe a genealogy upon which its present unity could be seen to rest. Though the Europeans of the fourteenth and fifteenth centuries were largely descended from the tribes that had swept into Europe at the time of the collapse of the Roman Empire, and until the Renaissance they had lost touch with the ancient world, they eagerly expropriated the Classical past for themselves. A popular way to understand the new worlds that Europeans encountered from Columbus on was to assimilate them to Europe's own unitary past – to the movement from pagan and barbarian to Christian and civilized. It was Rome, above all, that provided the political model and language for the developing European empires of Spain, Britain and France. This 'imaginative dependence of the new upon the old' (Pagden 1994, 12) helped define what an empire might be and also provided a common set of understandings around which European imperial competition was organized. Each of the European empires could see itself as the inheritor of the imperial mantle of Rome, at the same time as their common competition restricted the main players to Europe. The imitation of Rome (if not of Christ) was possible only within the boundaries of Europe. The classic division between Europe and Asia reaching back to the Greeks, therefore, was recycled to give the concept of Europe a degree of independence from that of Christendom.

Renaissance political theories reinforced this division between Europe, Africa and Asia. One was in the invention of the concept of 'balance of power', which was held to apply, from the time of the Florentine political theorist Niccolò Machiavelli (1469–1527) onwards, only to the balance between states in Europe. Another was in the total opposition drawn between European political practices and those elsewhere. In particular, Asia or the 'East' was seen as despotic and lacking in pluralist forms of political organization – this despite the historical record, as recounted by Patricia Springborg (1992), of a much more entrepreneurial and ungovernable East compared to a quiescent West lacking in much popular participation in politics until the twentieth century (also see Chapter 3). This historical inversion, according to Springborg (ibid., 20), can be understood as 'staking out territory and hoping, thereby, to create facts'. This point is also made somewhat more generally by Edward Said (1978, 54) in his book *Orientalism*, when he argues that Europeans defined themselves (as do all groups, one might add) negatively, against others whose nature was largely unknown:

> the sense in which someone feels himself to be not-foreign is based on a very unrigorous idea of what is 'out there', beyond one's own territory. All kinds of suppositions, associations, and fictions appear to crowd the space outside one's own.

The World of the East that Said identifies as crucial in defining the most important Other for the creation of Western self-images was in fact little known. This just goes to prove that a geopolitical imagination need not have much empirical content (Figure 2.2). But, crucially, as the historian Jerry Brotton (1998) has shown, it also meant denying the contributions of the East (from Christianity itself to the intellectual, artistic and navigational innovations of the Arab and Ottoman worlds) to European civilization. Finally, I would emphasize that it also required sweeping the rather variegated history of Europe since the collapse of the Roman Empire into a continental mould. In other words, Orientalism entailed a parallel Occidentalism that created a unified image of Europe at the same time that the non-European world was subjected to oversimplification and alienation.

Christianity played a continuing role in the self-image of Europeans but by the eighteenth century it was no longer the dominant force it once had been. By then the growing perception of Europe as a centre of artistic and intellectual invention had combined with the palpable sense of material progress to produce an awareness of Europe as a new civilization. This did not lead simply to a growing feeling of European superiority. It also engendered the idea of 'levels' of civilization with Europe at the top but with the possibility that other regions might achieve greatness by following in Europe's footsteps. Other places and people could be understood satisfactorily only in relation to this global cultural hierarchy. The one big division, between East and West, led to smaller ones, particularly as local differences were discovered in the course of exploration, trade and conquest. A whole fund of taxonomic lore developed, separating races, regions and nations from one another in terms of their relations to the most abstract global distinction. Even within Europe, the distinction between Eastern and Western Europe was important in delimiting where Europe 'ended' and Asia 'began'. Not only the most 'exotic' regions, such as the South Pacific or Brazil, provided grist to the intellectual mill of differentiating Europe from the Rest. Racial and cultural classifications drawn from Eastern Europe were, if anything, more important, given the centrality of the Europe–Asia divide to all other geographical divisions.

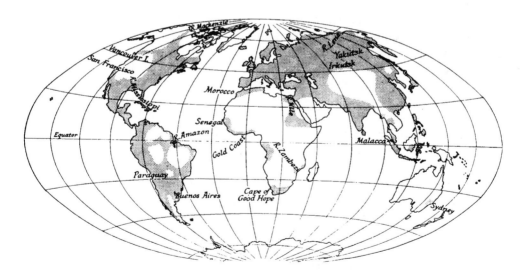

Figure 2.2 The world as known to Europeans, *c.* 1800. Shaded areas represent those parts of the world known 'in some detail' by that date.

Source: Crone (1978)

The idea of race was vital to the substantive distinctions drawn between different parts of the world. There is little or no evidence of a conscious idea of 'race' before the sixteenth century. Its high point was not reached until the late nineteenth and early twentieth centuries, when the fine distinctions of scientific racism replaced older, more informal means of drawing racial boundaries. In place of the uncertain stories about Shem, Ham and Japheth, humankind was ordered after the 1680s according to differences of anatomy, climate and pigmentation. Ironically, at first the old biblical division reappeared in which Northerlings, Middlers and Southerlings substituted for the three sons of Noah, if now on natural rather than theological grounds. Later taxonomies departed from the tripartite formula, tying physical appearance to a large number of cultural characteristics. The different races, however, continued to be clearly associated with different world regions and with different stages of economic and moral development.

Within the broad racial divisions of humanity, finer racial differences were identified that served to explain national origins in racial terms. For example, the French philosopher Montesquieu (1689–1755) remade the reputation of the northern barbarians, who heretofore had been thought of as the brutal destroyers of the Roman Empire, by conceiving of them as one of the 'races' of time and order, rooted over centuries in climate, land and soil. This served, and has served down the years since, to legitimize 'France' in a racial past independent of that of Rome. Indeed, disputes over this claim, and the counter-claim that France's origins lie in an older but derivative Gallo-Roman world, still animate French politics. In the autumn of 1996, for example, the Pope's arrival in France to celebrate the fifteen-hundredth anniversary of the Christian baptism of Clovis, the King of the Franks, resurrected old arguments about French racial origins, including claims that Clovis either was or was not the 'first Frenchman'.

The linkage of the local into global and imperial geographical taxonomies was facilitated by the differentiation of territories by their function within a growing global division of labour. This was increasingly thought of and justified in terms of the climatic and other environmental differences between places. Natural histories written in the nineteenth century portrayed the world increasingly in terms of North versus South rather than in terms of East versus West. Previously, Africa had not figured much in the emerging geopolitical imagination. This was partly because it was so unknown and thought of as basically inhospitable to Europeans. It had been something of an empty space rather than fundamental to European mapping of the world. But as a shift from civilizational judgement to scientific argument took place in the nineteenth century, Africa, the Americas, Asia and Europe had all to be considered in the great global horizontal scheme of things. The fundamental assertion was made that tropical climates encouraged idleness whereas temperate climates (such as those of Europe!) stimulated industry and innovation. This naturalization of differences in economic activity failed to note, of course, how much manufacturing and other 'advanced' activities were prevalent in climatic zones other than the temperate until killed off by European mass production and the subordination of tropical markets by temperate ones, owing little or nothing to any intrinsic character of those zones.

The attraction of this characterization to Europeans, however, is illustrated by the extent to which even late eighteenth-century British accounts of both economic development and cultural differences worked around differences, if yet crude, between frigid, temperate and torrid zones (particularly the latter two). This sometimes took powerfully gendered forms when climate was associated with differing incidences of sexual desire and the varying status of women was tied into the fundamental civilizational differences between the different zones. To the Scottish writer Adam Ferguson (writing in 1767), for example, hot climates were zones of unrestrained passion while temperate climates produced greater control. More particularly, European women were seen by many writers as enjoying privileges denied their sisters elsewhere. This both exalted the position of European women over that of women elsewhere and put differences between places in sexual behaviour and a wide range of associated cultural practices into the category of natural phenomena. Women in different postures and states of dress/undress came to symbolize the various regions of the world. If a crowned Britannia (in an eighteenth-century British publication) stood in for Europe, surrounded by attendants America, the East and Africa (Plate 2.2), Africa was often a naked woman and the East a beautifully draped courtesan surrounded by admiring men (see Nussbaum 1995, 67–69).

Plate 2.2 The Crowned Britannia, a frontispiece by G. Child in *A New Collection of Voyages and Travels*, vol. I (London 1745–47). The crowned Britannia, representing Europe, is seated holding her sceptre, and flanked by her attendants America, the East and Africa.

Source: The British Library

Through the frequent linking of luxuries with the feminine in eighteenth-century writing, the naturalization of difference by climatic zone also associated the tropical realms from which the bulk of the new luxuries came with the feminine. This was the global counterpart to the pervasive fear of the feminization of culture and politics among European political and intellectual élites of the eighteenth century:

> As wealth came more and more from the colonies and the stock market, and as cultivation at home became more efficient, the defenders of a ruralist (and masculinist) political ideal became more and more defensive. Power was seen as moving from country to city, and what came back to the country was the iconography of surplus wealth: parks and mansions occupied for half the year by stock-market millionaires, feminized personalities. ... The increased concentration of surplus wealth in the hands of an aristocracy and burgeoning middle class was seen as replacing an economy of need (masculinist subsistence) by an economy of desire (feminized superfluity).
>
> (Simpson 1996, 101)

The naturalization of the binary division of the world reflected in these oppositions reached its zenith in the geopolitical schemas proposed in the late nineteenth and early twentieth centuries. These played off the opposition between sea powers and land powers first noted by the ancient Greek writer Thucydides in his *Peloponnesian War*. They expanded the geographical scope from the case of Athens versus Sparta to that of maritime trading empires (such as Britain) versus autarkic land powers (such as Germany and Russia). A European classical education finally paid off in contemporary relevance. Geographical location in relation to the oceans and land masses suggested a kind of geographical fate against which agency was largely futile, save to point out the possibilities or options available to different parties depending on their relative location (Mackinder's formal geopolitical model of 1904: Figure 2.3). Historical analogy played an important part in making a model such as this commonsensical; putting it beyond critical scrutiny. For example, his reference to the history of barbarian invasions of Europe from Asia gave credibility to Mackinder's association of the Central Asian Heartland with threat and danger. There had been danger from there in the past. The geopolitical destiny this implied provided a 'binding in time' or 'a futureness to the past and a pastness to the future that is fundamentally reassuring' (Crocker 1977, 37–39). The formal geopolitical models (and similar general theories reducing local differences to global physical and environmental contrasts) therefore helped to stabilize the sense of a shrinking world in which rapid change was the order of the day. Though it is certainly arguable whether the formal models of Mackinder and others had that much direct influence on national policies, save perhaps later in Nazi Germany, much of what they preached was only a geographical formalization of contemporary common sense about global divisions based on the relative location of the continents, the causal effects of race and environment on the worldwide distribution of power, and the threats that each of the Great Powers posed to the others. By the time they wrote, much of the world had been incorporated into the European-based state system (including its extensions based on the territorial empires of the day). Even outsider states such as Japan and the United States now aspired to follow the imperial path laid down by the Europeans. The late nineteenth century also saw dramatic shifts in space–time organization with the spread of railways, telephones, steamship lines, foreign reporting, photography and cinema, and trade and investment flows. In a world of economic and social upheaval, geopolitical models offered an appealing stability and certainty.

They also offered a view of the world that could claim grounding in the natural features of the Earth. They did not express 'mere opinion'. They told it as it was. That they did so in the context of advising *different* states of their global prospects did not initially appear as contradictory. Only when a revisionist Great Power, such as Nazi Germany, appeared to have succumbed to the demonic influence of a spate of geo-political model building did the idea of a formal geopolitical model beyond the inter-ests of a particular state finally seem problematic. Not surprisingly, therefore, by dint of their association with the crimes of the Nazi regime in Germany, formal geopolitical models tended to lose favour after the Second World War.

The US containment and domino-effect doctrines of the Cold War (referred to in Chapter 1 in relation to the Vietnam War and later in Chapter 6) had echoes of the older geopolitical models, such as Mackinder's Heartland model, even though they lacked the formal grounding in the Big Picture of global geopolitics that these models offered. The visual metaphor of the domino effect proved particularly compelling. As the historian Frank Ninkovich (1994, xvi) has convincingly shown, from its origins in Woodrow Wilson's argument for US entry into the First World War down to the Vietnam War and beyond, the 'figurative image of toppling dominoes' tied the United States into a 'tightly linked system' which was not 'self-regulating' but required the application of

Figure 2.3 Mackinder's geopolitical model (1904). Halford Mackinder (1861–1947) was a major advocate of geography as 'an aid to statecraft', as well as a Conservative MP for a number of years. He revised his geopolitical model on a number of occasions, although its basic geographical structure remained the same.

Source: Redrawn from Mackinder (1904)

force to prevent the 'chain reaction' from reaching the 'last domino': the United States itself. Threats from a despotic and dangerous East (represented by the Soviet Union and 'Red' China) could be contained only by defending the first dominoes; otherwise all would be lost, as one country after another fell to the foe in a process of contagious diffusion. In another version the process of contagion was held to be analogous to the impact of one rotten piece of fruit on surrounding ones. A Manichean (total difference in religious status between entities) religious opposition sometimes entered into the domino calculus. Hell or Satan was seen to reside in the East (as it/he had resided in the medieval underworld described so imaginatively by the great medieval Florentine poet Dante). The satanic Other, or 'evil empire', to use the US President Ronald Reagan's characterization of the Soviet Union, was the uniquely wicked source of all travails facing humankind, working from the weakest towards the strongest links in the armour of the righteous rather than confronting the strongest outright. Soviet leaders used similar representations (relying more on metaphors of organic growth and decline akin to the 'rotting fruit' version of the domino theory than on religious imagery) in their accounts of the great global struggle between an embattled socialist experiment and an aggressive but decadent world capitalism.

By the mid-twentieth century, however, a number of dominant spatial practices and representations had underwritten the continuing connection between local differences and the binary division of the world, largely independent of formal geopolitical models. One was the growing interconnection between parts of the world as colonial economies generated regional specializations (such as cotton in the US South and rubber plantations in British Malaya and the Dutch East Indies) that worked as components of a multilateral trading system joining together Britain, other industrializing countries and the colonies in a global system of production and exchange. The colonized South was assimilated into the place of the ancient East; but with much the same stereotypes justifying its subordination. In the years after the Second World War the Mother Country–Colony distinction that had reproduced the East–West division for so long under colonial conditions broke down under the onslaught of decolonization movements and the emergence of the Cold War conflict between the United States and the Soviet Union. This gave way in the 1950s to the Three Worlds concept of a normal, natural West (the First World) challenged by an unnatural, state-dominated East (the Second World), with each vying with the other to produce political–economic disciples in a Third World of traditional yet developing countries that became the typical representation of the enduring geopolitical division between East and West. The governments of the United States and the Soviet Union represented themselves and one another in ideal-typical terms as global agents engaged in a struggle over the heritage of the West. The Soviet Union proved especially vulnerable to the charge of oriental despotism, even though its political economy and system of governance had good Western credentials. It became the wicked Other against which the identity of the United States was defined. Much domestic as well as foreign policy in the United States was made in response to perceptions of the threat that the Soviet Union posed to the US view of world order and vice versa. What is most important is that a binary geography of East versus West again dominated the understanding of how the world works. The category of the Third World derived from the structural opposition drawn between the other two worlds: an American-allied First World confronting a Soviet-dominated Second World. The regions and countries within the vast zone of the Third World, therefore, were reduced to the role they performed in the conflict between the other two rather than being meaningful in and of themselves (see Chapters 3 and 6).

The militarized nature of the conflict between the United States and the Soviet Union served to further bifurcate global space. At one and the same time, the advent of the capacity to deliver nuclear weapons over great distances almost instantaneously both devalued the military importance of territorial space through a new emphasis on virtuality and yet reinforced the sense of being targeted because of where you happened to live. The net outcome was an enhanced sense of belonging to a targeted space, which in Cold War terms meant being either of the East or of the West. You had to be on one side or the other, given the targeting patterns of the weapon systems. Even as deterrence based on open knowledge of the other side's destructive capabilities was replaced by the dissimulation implicit in decoy, STEALTH and false radar signature technologies, 'home' still depended on which side you were on. For example, 'duck and cover', the leitmotif of US civil defence programmes in the 1950s (imagine schoolchildren scrambling under desks), created the illusion that it was possible to survive a nuclear attack and live to perpetuate the values of the sanctuary, the American home, that such an attack would necessarily violate.

Since the end of the Cold War in the early 1990s the Three Worlds concept has lost much of its appeal. In addition, growing disparities in economic development within the Third World category have made its use alone without reference to the other two increasingly problematic. The disappearance of an immediate nuclear threat has likewise removed that element in the East–West opposition that rested on making a totalistic military distinction between the First and the Second Worlds. The emerging binary divisions seem to be either those drawn between the information rich and poor, relative to access to global telecommunication networks and flows of information (largely in English) about finance and production, or those positing a continuing cultural East–West divide with respect to 'civilizational' conflicts emerging along global cultural fault lines (particularly that of the 'West versus the rest'). The first of these suggests a much patchier and more localized pattern of global difference whereas the latter is indicative of a continuing global division to which local differences must be assimilated. The struggle between these two visions and the degree to which spatial practices are organized around each of them will be important (along with attempts at reinstating state-centredness) in determining what, if any, prospects the modern geopolitical imagination will have for reconstruction in the future. I will return to these two competing visions in Chapter 7.

Conclusion

World politics has developed to its present form because of the way in which the world itself has been brought into focus over the past five hundred years. In brief compass, I have shown how two aspects of global visualization have produced the major geopolitical taxonomies around which world politics is organized. Seeing the world as a structured 'whole' required reaching back to older cosmographies to find a relevant model for understanding the world that Europeans started to encounter from the late fifteenth to the early sixteenth centuries onwards. That of Ptolemy provided a particularly auspicious beginning because of its openness to discovery and the filling in of new information. The rapidly emerging linear perspective associated with the architectural and cartographic practices of the Renaissance added a critical element: the separation of the observer from an object under vision and the object's location within a visual field that would determine the meaning it would take. This had the effect of allowing the world to appear as a meaningful whole in which any part of it could be understood only in relation to the whole. It also led to an emphasis on the self-evident reliability of information as it appeared on maps of the whole. Maps were guides to territory, not templates framing the selection, naming and ordering of places. These conceptual innovations provided the backdrop for a more specific and recurring tendency to divide the world into two opposing zones, each of which defined the other by communicating what it is not. The most important of these has been a persisting West–East divide (subsuming a North–South divide). Though reproduced in a number of manifestations, depending on the current mix of material conditions and related representations (see Chapter 6), this division has ancient roots in Western thought. But it also works with and confirms other basic oppositions between temperate and torrid climates, pluralist West and despotic East, sea powers and land powers, and civilized and barbarian, that relate ideas about the nature of economic and political practices to geopolitical visions of how the world works as a whole. Such essential oppositions define the horizons of the modern geopolitical imagination as developed from the late fifteenth and the sixteenth centuries to the present. The most typical of these within the modern geopolitical imagination, that of primitive or backward versus modern, is the subject of Chapter 3.

Recommended reading

Brotton, J. 1998 *Trading Territories: Mapping the Early Modern World*. Ithaca, NY: Cornell University Press.

Cosgrove, D. 2001 *Apollo's Eye: A Cartographic Genealogy of the Earth in the Western Imagination*. Baltimore: Johns Hopkins University Press.

Eco, U. 1999 *Serendipities: Language and Lunacy*. New York: Columbia University Press.

Grafton, A. 1992 *New Worlds, Ancient Texts: The Power of Tradition and the Shock of Discovery*. Cambridge, MA: Belknap Press of the Harvard University Press.

Gregory, D. 1994 *Geographical Imaginations*. Oxford: Blackwell.

Gress, D. 1998 *From Plato to NATO: The Idea of the West and Its Opponents*. New York: Free Press.

Harley, J. B. 1989 Deconstructing the map. *Cartographica*, 26: 1–20.

Helms, M. 1988 *Ulysses Sail: An Ethnographic Odyssey of Power, Knowledge, and Geographical Distance*. Princeton, NJ: Princeton University Press.

Lestringant, F. 1994 *Mapping the Renaissance World: The Geographical Imagination in the Age of Discovery*. Berkeley and Los Angeles: University of California Press.

Lewis, M. and Wigen, K. 1997 *The Myth of Continents*. Berkeley and Los Angeles: University of California Press.

Nussbaum, F. 1995 *Torrid Zones: Maternity, Sexuality, and Empire in Eighteenth-Century English Narratives*. Baltimore: Johns Hopkins University Press.

Pagden, A. 1994 *Lords of All the World: Ideologies of Empire in Spain, Britain and France, c. 1500–c. 1800*. New Haven, CT: Yale University Press.

Roberts, D. 2001 *Great Exploration Hoaxes*. New York: Modern Library.

Whitfield, P. 1998 *New Found Lands: Maps in the History of Exploration*. New York: Routledge.

Wolf, E. 1982 *Europe and the People without History*. Berkeley and Los Angeles: University of California Press.

Zerubavel, E. 1992 *Terra Cognita: The Mental Discovery of America*. New Brunswick, NJ: Rutgers University Press.

3 Turning time into space

In this chapter I want to suggest that a second element of the modern geopolitical imagination has involved translating time into space. In other words, the blocks of space identified in Chapter 2 have been understood in terms of the essential attributes of different time periods relative to the idealized historical experience of one of the blocks: Europe or, more broadly, the West. Hence, territories are named as primitive versus advanced or backward versus modern in relation to an idealized version of European experience.

Numerous critiques of social science thinking about social change have noted this move. My purpose in this chapter is to extend the critique in two ways: first, to argue that converting time into space, so to speak, is a fundamental characteristic of modern geopolitical thought; and second, to claim that this conception of space pervades contemporary writing about national development irrespective of where in the world the state in question is located. Even though this way of thinking developed along with European colonialism, it cannot be reduced to it, in the sense of applied only to those parts of the world on the receiving end of European empire building. Rather, the whole world has had to fit into this conception of social change, not only those parts outside the European realm but subject to direct European economic and administrative control, but Europe itself, too. Not surprisingly, therefore, the mode of understanding has continued to survive long after formal colonialism has officially ended. Imperialism, in the sense of the understandings and practices of the modern geopolitical imagination stimulated by early encounters outside Europe (and its extensions, such as the United States), lingers on even after the demise of colonialism. To illustrate the general argument about the pervasiveness of the expression of time in spatial terms I use the examples of Australia and Italy. These show that the division of space in terms of time sequences is not just a simple global one, as illustrated perhaps by the Australian example, but has become both more abstract and generic in its applications, as illustrated by the way Italy is conceived of in much contemporary thinking.

Mapping the unknown

Why does this matter? For one thing, understanding of both time and space has suffered as a consequence of expressing one in terms of the other. In particular, time has lost its dynamic character as it has been reduced to a twofold categorization of space. As a result, space has also been oversimplified and reduced to a set of simple containers for either/or statements about its characteristics. The language of modern thought is filled with the fusion of time with space into binary distinctions between those areas ahead or advanced and those areas behind or backward. Indeed, one of the main organizing devices for thinking about the world, the intellectual division of labour between specialists in the 'Three Worlds of Development' – students of the advanced capitalist world, students of the (former) Communist Second World and students of the barriers to development in the Third World of former colonized peoples – derives from the imposition of an idealized version of the past development of the First World on to the spaces occupied by the two other worlds.

The juxtaposing of the temporal stage with the spatial category is vital to the modern geopolitical imagination because it provides a natural link between the European past, on the one hand, and the global present outside of the modern world, on the other, in terms of what the latter lacks and what the former has to offer to make up for this deficiency. The projection of temporal qualities drawn from a rendering of a specific historical experience on to terrestrial space enables three political–intellectual positions. One is the tendency to *essentialize* places, or identify one trait as characterizing a particular spatial unit (e.g. caste in India, the Mafia or political instability in Italy). A second is the temptation to *exoticize*, or focus on differences as the single criterion for comparison between areas. Similarities or universal conundrums are thereby eliminated from consideration (e.g. difficulties of social mobility everywhere, general barriers to political participation). The third is the tendency to *totalize* the comparison, or turn relative differences into absolute ones. The whole of a society is thereby made entirely recognizable in any one of its parts. The whole block of space is suffused with the character that is defined by the social totality (usually a 'culture' understood as a spatially indivisible unity). These have always been questionable modes of thinking. In the contemporary world of global flows of people, goods and money they are all too problematic.

The problem under consideration is not just a theoretical one. The relative adequacy of a particular account of how the world is divided up and characterized so as to make the best sense of it is also a normative problem. Places and people are understood not on their own terms but only in so far as they slot into the global scheme of things as spaces that are at this or that 'stage' of development relative to an idealized European past. The particularities and peculiarities of Europe and elsewhere are thus swept into global categories based solely on Europe's presumed more advanced status. Whether this must remain for ever the case, however, has been more contentious. In fact, by the eighteenth century the idea of a permanent opposition between advanced and primitive started to give way to the idea that the primitive could advance in status. This belief has been the leitmotif of modern doctrines of economic development, whether of American capitalist or Soviet Communist provenance.

The sources of this approach to mapping the unknown by converting time into space are twofold. Adopting Ptolemy and other ancient sources as cartographic guides to the world as a whole, even if they were regarded as elementary ones, had fateful consequences. It reinforced the tendency to think of Europe as the centre of the world and to elevate Europe's past, as symbolized by the work of Ptolemy and others, to the universal standard. Others elsewhere just could not measure up. But this sense of superiority was reinforced by the physical, bodily presence of Europeans around the world. Not only had the Europeans travelled further and faster than anyone had ever done previously, they also managed to overwhelm and conquer others because of the typically greater sophistication of their weapons, and to profit from their usually greater commercial and financial acumen. The 'unknown' regions of the world, therefore, notwithstanding the genuine interest they often elicited for their particularities, were already 'known' in terms of a global standard of excellence that had Europe as its model. The modern geopolitical imagination has simply inherited this conception of 'mapping the unknown'.

Australian journeys

Of course, the experience of European encounter with the rest of the world was always more complicated and less uniform than geopolitical categorization would lead one to suspect. The movement outwards from heart to edge of empire was uneven and produced diverse forms of colonialism. In circumstances such as those in India where outsiders confronted a densely populated and recognizably sophisticated civilization, the existing social order was largely maintained but reoriented to the economic interests of the imperial core. The colonization of Australia, however, was premised on the view that its indigenous society was primitive and its view of property conflicted with the needs of ordered settlement. This did not mean that all explorers and early settlers thought of the Australian interior as an unoccupied land freely available to all settlers. There were frequent disputes over Aboriginal land rights. There was a sense of wonder at the ability of Aborigines to survive and, to a degree, prosper in environmental conditions that Europeans saw as inhospitable. The outcome of imperial domination, therefore, was contested and contingent. But the stories told to later generations of 'new' Australians conformed much more to the heroic narrative of the 'modern' sweeping away the 'backward'. Explorers also tended to use similar heroic terms with which to characterize their own activities in confronting unknown and dangerous peoples. In every case, therefore, the explorers of Australia already knew how to interpret what they would find before they left the coast for the interior. The modern geopolitical imagination owes much of its appeal to such simplification of history: sorting empirical information into already operative sets of pigeon-holes.

The political and moral consequences of the spatial representations and practices involved in turning time into space are thus well illustrated by nineteenth-century European encounters with Australian Aboriginal peoples. To Europeans, the boundary between them and those they encountered was a line of proprietary appropriation, defining the limits of the modern in contact with the traditional. To the Aborigines, however, boundaries were 'debatable places' which they regarded as zones for inter-tribal communication. Of course, Australia was finally 'settled'. It became a settler colony with a largely non-indigenous population. As a result, the boundaries that emerged served not to demarcate a cultural encounter but to impose a European presence. This violent imposition,

> which substituted for conversation, [was] institutionalized in the form of what is represented as 'Australia' just as other names and boundaries on the dominant, geopolitical world map are rigidified and thus removed from the possibility of encounter. . . . It is time to unread the old map and begin the process of writing a new one, a process without limit.
>
> (Shapiro 1995, 40–41)

Australia can be described as a settler colony that managed to achieve the transfer of the imperial power that had given it its first existence as a geographical entity from its core in Britain to the colony itself. Consequently, Australia does not fit the category of the Third World in which imperial power remained largely alienated from its creations; but neither does it neatly fit into that of the European First World. Rather like the popular history of the United States for Americans, the history of Australia's exploration and settlement still bulks large in the self-images of Australians and in contemporary Australian politics. Recent political disputes over finally casting off the last ties with Britain by becoming a republic, and court cases over Aboriginal land claims, are reminders of the origins of the country as a settler colony (see Figure 3.1). The success of at least a measure of redress for the Aboriginal population in the form of land claims also suggests that Australians are finally coming to terms with what the Australian High Court called a 'national legacy of unutterable shame' in the historic relationship between European settlers and Australia's pre-conquest population (Pagden 2001, 170). Australian experience also suggests, however, that the terms of engagement established at the time of original exploration and settlement are not easily transcended. The nature of European engagement with a world outside Europe's own lives on today, therefore, in the imaginations and politics of many who had nothing to do with the original encounter.

Australia was 'opened up' to European exploitation and settlement by large numbers of adventurers, traders, missionaries, criminals on the run and more conventional 'explorers'. There was a considerable range of opinion about the native inhabitants, labelled Aborigines from an early point in the process of exploration. What seems clear, however, is that the terms of debate were already set before any exploration took place. Two strategies of understanding predominated in the accounts of nineteenth-

Figure 3.1 Aboriginal land claims in Australia (1997).

Source: Author

century explorers of Australia. One was to interpret native customs with respect to either classical European culture (that of the ancient Greeks and Romans) or relative to stereotyped views of the East (or Orient) that Europeans thought that they knew. The explorer Sir Thomas Mitchell (1839), for example, interpreted the Aboriginal use of green boughs of leaves upon meeting visitors as indicative of benign intentions. This involved explicit reference to the ancient Greek association of green boughs with peaceful intent. Plausible as this might be, it required the picturing of Aborigines with respect to the European past rather than to a different culture coexisting with them in time and space. On many occasions explorers compare what they think the Aborigines are up to with what they already know as exotic from the Arab, Chinese and Indian realms. This creation of the Aborigines as familiar, however, did not extend to regarding them in any way as equals of Europeans. A second strategy of understanding uses a number of devices to set the Aborigines apart as totally different from the explorers themselves. In the first place, a pre-existing language of race was used to classify the Aborigines in relation to a range of oppositions: white versus black, civilized versus savage, Christian versus heathen, rational versus instinctive, etc. The civilized versus savage opposition was particularly widely used to explain reactions to the explorers' presence. Thus, innate 'savage' urges were identified as explaining why Aborigines might object to the draining of water resources by a visiting explorer. That this kind of reaction was a rational response to the depredations visited by the intruder never seems to have occurred to the explorers. At the same time, the Aborigines were identified with a 'wilderness' in which rational definitions of individual ownership did not apply. This was put down to the fact that Aboriginal society was at the bottom of the global social hierarchy. This hierarchy associated peoples in the European past, such as people of the Bible, and the Orientals, against which European civilization had become defined, with the Aborigines. A pre-existing mindset brought into the field was used to interpret what was discovered.

These understandings mattered because explorers were in the vanguard of European settlement and provided a set of founding heroes for the emerging colony–nation of Australia. They recapitulated understandings that Europeans took with them wherever they went. Not that all said or thought exactly the same; it is more the case that there was a common semantic field of understanding using the same vocabulary and eliciting similar comparisons with the European past and the seemingly more familiar Orient (as described in Chapter 2). One explorer, George Grey (1841) (quoted in Ryan 1996, 158), captured both this historical equivalence and the sense of European superiority in relation to the present condition of the Aborigines when he wrote in his journal about Aborigines watching his camp from the distance:

> with the help of my telescope I once distinguished their dusky forms moving about in the bush. A large flight of cockatoos, which lay between us and them, were kept in a constant state of screaming anxiety from the movement of one or the other party, and at last found their position so unpleasant, that they evacuated it, and flew off to some more quiet roosting place. Their departure, however, was a serious loss to us, as they played somewhat the same part that the geese did in the [Roman] Capitol; for whenever our sable neighbours [Aborigines] made the slightest movement, the watchful sentinels of the cocka-toos instantly detected it, and by stretching out their crests, screaming . . . gave us a faithful intimation of every motion.

Not only does this description call up a classical reference familiar to British school-boys of that era, it also implies that the Aborigines are the transgressors, in danger of trespassing upon the explorer's land. That it is in fact the Aborigines who stand in literal succession to the Romans in this historical analogy is, of course, totally lost on the observer.

Not simply 'out there', Australia was created in the encounters of Europeans with the continent and its inhabitants. But the discovery was not just a naïve response to what was encountered. As elsewhere in the world colonized by Europeans, the encounter was actively mediated by existing ideas about the range of societies and their relative power and sophistication. From this point of view, Australia was already understood before it was discovered! And it was understood largely in terms of the stages of a European past imposed on everywhere else.

Strange Italy

The conceptual grid figuring spatial differences in temporal terms has become so univer-salized that even differences within Europe have also come to be thought of in the vocab-ulary of backwardness and modernity. So, even while on a global scale Italy is modern, now the world's fifth largest economy with an income per capita higher than that of Britain, within Europe it is often still seen as exhibiting clear signs of backwardness: political corruption and instability, organized crime, commitment to emotion more than reason, flair for decoration and presentation. As with the Australian case, a moral judge-ment is implied by this usage. The differences between it and its neighbours are *total-ized*. The position seems to be that there is something lacking about Italy in comparison with its neighbours that accounts for these differences, ergo it has lagged behind them on the path to modernity.

How common is the invocation of the backward–modern metaphor in relation to Italy? The answer is: very common indeed. Discussion of Italy's relation to modernity and modernization has been central to debate in studies of Italy by Italians and others. Adoption of the vocabulary of backward and modern is not restricted to any particular school or grouping but is common across the political spectrum. For example, the debate over the nature of Italian Fascism (1922–43) is dominated by claims and counter-claims about its modern and traditional qualities, irrespective of the political commitments of the authors. All leading genres of writing about recent Italian history are also dominated by this way of thinking. I want to focus here on two examples, one drawn from British writing about Italy (Paul Ginsborg's book (1990) on the history of post-war Italy, 1943–88) and one from the work of a famous Italian anthropologist (Carlo Tullio-Altan).

Ginsborg's detailed account of Italian history from 1943 to 1988 works around a series of oppositions through which movement from backwardness to modernity is inter-preted. In order of degree of abstraction these are weak state versus strong society, family orientation versus collective action, class and regional dualisms: workers versus bour-geoisie, north versus south, corrective versus structural reform, and militance versus *riflusso* (retreat into private life). Ginsborg operates from a standpoint sympathetic to the condition of peasants and workers in the early post-war period. He sees the period 1943–48 as one in which opportunities to redistribute wealth and power were lost, unlike in Britain, where important changes took place. Subsequent history is read in terms of the consequences of this fateful failure. But despite this failure, by the 1980s a nationalization of values had finally captured even the most traditional of Italians, though Ginsborg seems doubtful as to its permanence. This is because Italy's modernity is uniquely fragile, threatened by the persisting possibility of reversal of the current balance of oppositions. The path to modernity taken after 1948 is still subject to unpredictable shifts and pathological turns. Consequently, as Ginsborg concludes for Italy in the 1980s, Italy remains a country with 'a deformed relationship between citizen and state' (1990, 421), where 'Neither from civil society nor from the state has there emerged a new and less destructive formulation of the relationship between family and collectivity' (ibid., 418) and, more particularly, 'the lack of *fede pubblica* (civic trust) continues to bedevil southern society' (ibid., 417). Thus, for Italy, unlike Britain, true modernity is always around the corner, promised but never finally arriving.

From another perspective, however, it is not so much the absence of modernity as the continuing presence of backwardness that exercises attention. In his book Tullio-Altan (1986) sees Italy as condemned to perpetual repetition of a condition of civic immaturity. He attributes this to the temporal persistence and geographical spread within Italy (from south to north) of the syndrome of amoral familism first diagnosed by the American political scientist Edward Banfield in his book *The Moral Basis of a Backward Society* (1958). To Tullio-Altan, the roots of Italy's backwardness lie in the initial failure at the time of unification in the period 1860–70 to overcome the national dualism between a developing north and a backward south. He makes much of the prophecy of Mazzini (one of the leading figures in the movement for Italian unification) that 'Italy will be whatever the Mezzogiorno [southern Italy] will be' (Tullio-Altan 1986, 16; my translation). Two phenomena are identified as largely responsible for the 'southernization' of Italian society: *clientelismo* (the exchange of votes for favours) and the practice of *trasformismo* (collaboration among politicians by the exchange of favours). The spread of these practices has prevented the development of a national ruling class with a truly national orientation. By its continuing geographical fragmentation Italy has been trapped in a condition of cultural backwardness compared to more successful national societies.

The main claim is that the Italian political class embodies both a culture that is particularistic and a parallel political practice that is based on the exchange of favours. The presumption is that these are a straightforward inheritance from the past. But what if they are new rather than old? What if they are elements of a 'modern and effective system of power designed to integrate into the national society masses of people who are dangerously inclined to claim democratic participation and their own emancipation' – in other words, 'a modern political culture' (Signorelli 1986, 45)? Furthermore, the ideal of civic maturity is hardly realized anywhere on earth, so why hold Italy to such an impossible standard? The discouraging of political participation and the subverting of political institutions for personal gain are features of most political societies. They are definitely not unique to Italy.

The attribution of backwardness makes sense, therefore, only if the backward features of Italy are implicitly compared to some ideal of modernity. For Tullio-Altan these would obviously include the absence of the vote–exchange nexus that he gives so much attention to in his account of Italian backwardness. But he also offers some clues as to the more positive features of modernity. He draws a stark opposition between societies organized with reference to a social conscience based on individual responsibility, which he sees as produced by the Calvinist branch of the Protestant Reformation, and those, such as Italy, in which an 'amoral familism' prevails. This is Banfield's idea that a modern conscience fails to develop when people's moral horizons are defined by the bounds of a nuclear family. Italy's backwardness, therefore, derives ultimately from the failure to have experienced the Protestant Reformation.

In fact, the Italian case indicates something of the arbitrariness attached to attributions of backwardness and bolsters the suspicion that they may be political in origin. For example, there is much evidence that Italy has long had a more compact and nationally oriented political élite than many other European nation-states, including France and Germany. Also, if one focused on technology, industrialization or consumerism as attributes of modernity, then Italy could hardly be construed as backward. The imprecision of the terms 'backward' and 'modern' gives them a mysterious quality that allows them to be used to paint a picture using whatever materials are suitable to make the particular case. In this case the purpose seems to be identifying the South as the source of Italy's recent political woes and labelling as indicators of backwardness those features of Italian politics and society that are most strongly associated with that region.

Given this arbitrariness, the question inevitably arises: why should attributions of backwardness and modernity have such appeal? In the Italian case two factors stand out. The first is that there is a widespread sense on the part of Italian intellectuals of the failure of Italy to 'mature' as a political entity. This is not an imposition by outsiders, as an attractively simple innocent insiders versus evil outsiders account might have it. This view has deep roots in Italy itself. For example, the main theme of the great Florentine writer Machiavelli (1469–1527) in such Renaissance-era works as *The Prince* and *The History of Florence* was the civic corruption of his epoch compared to the civic excellence of the ancient (republican) Romans. The power of the popes and intervention in Italy from outside were indicted as the main culprits. Again, the main figures in achieving Italian unification in the nineteenth century blamed the reactionary Austrian and Bourbon (Spanish) regimes that governed large parts of the peninsula and islands for the social and economic decline that Italy experienced after the Renaissance (Figure 3.2).

In the second place, Italy has also been plagued ever since unification by a disunity based on geo-economic and linguistic fragmentation. The city-states so closely associated with the European Renaissance (Florence, Venice, etc.) failed to produce an integrated state that would have led Italy towards the model followed by France, Spain and England. The variety and density of cities with their own hinterlands worked against the ready creation of a territorial state with a dominant capital city. Foreign rulers found in Italy propitious circumstances for a strategy of divide and rule. The continuing secular power of the Roman (but Catholic) Church has continually frustrated attempts at creating a national state in its homeland. As a result, Italians have had to look outside of Italy for inspiration, to the extent that 'admiration for foreigners has been the principal ingredient of Italian nationalism' (Lanaro 1989, 212; my translation)!

Spaces of backwardness

One consequence of Columbus's return from his famous voyage of 1492 was a heightened sense among European intellectuals of a hierarchy of human societies from primitive to modern. It is surely no coincidence that in conventional accounts of the periods of history, 'modern' history begins with the era of Columbus. However, the simple juxta-position of new worlds against a familiar old world from which the discoverers came is an altogether too simplistic view of what happened. All societies define geographical boundaries between themselves and others. Sometimes the world beyond the horizon is threatening, sometimes it is enticing. But not all societies engage in portrayals of the other societies they come across as 'primitive' or 'backward'.

In fact, little is known of how exactly early modern Europeans assimilated exotic peoples into their understandings of historical geography. In a major review of schol-arship covering the sixteenth and seventeenth centuries, however, Michael Ryan, for example, suggests that the major means was through the assimilation of the exotic into Europe's own pagan and savage past. 'In the triangular relationship among Europe, its own pagan past, and the exotic, the principal linkage was between Europe and antiq-uity' (Ryan 1981, 437). The categories of 'pagan' and 'barbarian', discovered as an inheritance from the European ancients, were deployed to differentiate the newly discov-ered worlds from the old one. Thus, a conception of the temporal transition through which the European social order had been transformed was imposed upon the spatial

Figure 3.2 The political geography of Italian unification, 1860–70.

Source: Author

relationship between the new worlds and Europe in its entirety. The religious dimension was especially important in reading the new pagan worlds as standing in a relation to the (European) Christian world as that world stood in relation to its own pagan past.

This is not entirely surprising if it is remembered that the first discovery by Europeans of the new geographical worlds coincided with the rediscovery of Europe's own ancient past in the European Renaissance (noted in Chapter 2). Indeed, as Mandrou (1978, 17) recounts,

> The new worlds that fascinated the intellectuals of the sixteenth century were not so much the Indies – West or even East – but those ancient worlds which the study and comparison of long-forgotten texts kept revealing as having been richer and more complex than had been supposed.

Of course, Italy was the centre for this new activity, associated as it was with monastic libraries, universities and the recovery of ancient texts. Ironically, given the challenge to ecclesiastical authority that the rediscovery of the ancients could entail, it was from within the Church that the new learning arose: 'Italian cities, richer in Churchmen than any others in Europe, and closer to the papal authority, constituted the setting that was most apt to stimulate the study of ancient texts and pre-Christian thought' (ibid., 23).

The articulation of spatial differences in temporal terms was reinforced by the taxonomic lore that Renaissance-era Europeans learned from the ancients. As Edward Said (1978, 57) has suggested, much ancient Greek drama involved demarcating an 'imaginative geography' in which Europe and Asia are rigidly separated. 'Europe is powerful and articulate; Asia is defeated and distant. . . . Rationality is undermined by Eastern excesses, those mysteriously attractive opposites to what seem to be normal values.' An image of essential difference with roots sunk deeply in the primordial past was used to invent a geography that had no real empirical points of reference. In terms of such categories as race, property, oligarchy and economy, the Orient (and non-Europe in general) was claimed as 'the negation of all that was being claimed for the West, by polemicists knowing, in fact, very little about it' (Springborg 1992, 20). From this stereotyping derive such views as the association of democracy with the West and despotism with the East, history with the West and cultural stasis with the East, ingenuity with the West and conformity with the East. These views never rested on empirical evidence but rather on a-priori assertion. Not only did 'West' and 'East' have long histories of interaction and interchange; Christianity itself came from the East, but the histories of each were complex and multifaceted. Neither was the historic 'home' of either democracy or despotism. The reduction of cultural complexity to a simple two-region opposition had discrete historical origins (also see Chapter 2). The 'Ottoman peril' (threat of Turkish invasion) of the Renaissance period gave particular credibility to the sense on the part of vulnerable Europeans of a profound chasm between the familiar world of Europe and the exotic world of the Oriental Other. In Europe 'The Turkish threat worked toward reviving a waning loyalty to the *Respublica Christiana* [the idea of Europe as Christendom] and gave new life to the old cry for peace and unity in a Christendom subject to the pope' (Schwoebel 1967, 23). However far from Europe Europeans might venture, this deep sense of the civilized versus primitive and Christian versus heathen travelled with them. The explorers and settlers of Australia, therefore, could be expected to arrive there with prior expectations about the others they might meet. They certainly did not show up without a vivid sense of the differences that separated them from their more immediate neighbours such as Turks, Moors and Arabs.

As the European states emerged from the dynastic struggles and religious wars of the seventeenth century and embarked upon their schemes of empire building outside Europe, comparisons of themselves to the ancient world, especially the model provided by Rome, proved irresistible. Lord Lugard (1926, 618), the British ruler of northern Nigeria, was to maintain towards the end of the colonial period that Britain stood in a kind of apostolic succession of empire: 'as Roman imperialism . . . led the wild barbarians of these islands along the path of progress, so in Africa today we are re-paying the debt, and bringing to the dark places of the earth . . . the torch of culture and progress'. Perhaps the most important exponent of this Roman-inspired view of history was the German philosopher Hegel (1770–1831), especially in his *Philosophy of Right*, first published in 1821. On the basis of the relative extent of the absolute sovereignty of the state and what he called its 'ethical substance', the nation, Hegel divided the world into four historical realms arranged hierarchically, with the Oriental (India seems to have figured prominently in his thinking) as the lowest, the Germanic as the highest (surprise!), and the ancient Greek and Roman worlds, as the precursors of the Germanic, in between. In Hegel's understanding (which fits in so well with that of later generations of Italians), the achievement of an integrated nation-state was a necessary condition for achieving moral identity (see Chapter 4). From this point of view, Hegel is particularly important as a philosopher of the geopolitical imagination. He saw the world as a whole, dividing it up into geographic chunks on the basis of levels of political development vis-à-vis an interpretation of European experience of which the most important part was the European nation-state, in the process of replacing a greater variety of forms of political organization (dynastic empires, papal states, feudal territories, systems of shared power such as the Holy Roman Empire, etc.) at the time he was writing.

Beginning in the late eighteenth century the resort to classical precedent to understand spatial differences in the social order was put on a scientific foundation. Pragmatic common sense was backed up by explanation in terms of natural processes or by analogy to natural processes. It became increasingly popular to see social change as a transition from one stage or level of development to another. As the nineteenth century wore on, and in imitation of discourse within biology, this view was elaborated upon as an evolutionary movement from a lower to a higher level of organization. This trend took two directions. One direction was that 'race' shifted in meaning from the looser sense of groups differing as much civilizationally as physically to a much stronger sense of complexion and physical stature as determining civilizational differences. If an earlier generation of Europeans, including many of the intellectuals of the Enlightenment such as Kant (1724–1804), had used notions of racial difference to emphasize their break with the 'Dark Ages' in which the rest of the world was still living, by the mid-nineteenth century race had taken on, in the writings of such anti-Enlightenment figures as Gobineau and Le Bon, the connotation of the determining factor in explaining human history. A second direction was that some parts of the world were increasingly seen in terms of levels of development – that is, levels of economic growth and social and political progress – that Europe had experienced previously. But this difference no longer need condemn the benighted to perpetual backwardness. If the experience of the 'developed' were imitated, then those that lagged behind could begin to catch up. The idiom of what the Indian historian Ranajit Guha (1989, 287) terms 'Improvement' came to prevail over that of 'Order'. The distinction no longer lay primarily in essential difference that could not be transcended but in the possibility of overcoming backwardness through imitation. The future of the backward lay in repeating what Europe had done.

The power of this latter idea is indicated by the degree to which political élites around the world have enthusiastically adopted the ideal of modernity as a recapitulation of European and, more recently, American experience. For example, even the apparently anti-Western governments of China since 1949 have used a rhetoric appealing to a catching up or overtaking of Europe. At the height of the Maoist period in the late 1950s, when Western ideas were supposedly anathema to the regime, the economic plan of the time, the Great Leap Forward, invoked the appeal to *chao ying gan mei* (literally, 'overtaking England and catching up with the United States'). This reflected a long-standing commitment of the Chinese élites (irrespective of political ideology) to the idea of *zhong xue wei ti, xi xue wei yong* ('Chinese learning for fundamental principle, Western learning for practical use'). The problem has been that it has proved difficult, if not impossible, to keep Western influence confined to the technological realm. At the same time, it has never been entirely clear what precisely is 'purely' Chinese to defend from outside assault. The tendency has been to assert a Chinese essentialism or total difference with everywhere else based, above all, on the difficulties of material existence in China. This mandates a narrow definition of human rights associated with feeding the people, even though there is nothing about the history of China that would necessarily lead to such a limited conception. There is nothing particularly Chinese about denying freedom of speech or trial by law.

By the close of the nineteenth century, however, modernity was increasingly conceived of within Europe as the form of society in which social interaction is rationally organized and self-regulating. The German economist/social historian Max Weber's (1864–1920) theory of rationalization provided the most important account of this modernity within an evolutionary conception of social change. For Weber, writing at the end of the nineteenth century, the rationalization of social life in the modern (Western) world involved the increasing regulation of conduct by instrumental rationality: using criteria of merit in employment, for example, rather than 'traditional' norms and values using caste, family and kinship. Weber himself was less than enthusiastic about this process of modernization. He associated modernity with the rise of bureaucracy and impersonal regulation of social behaviour. But many of his sociological disciples have had few doubts. The version of Weber's theory disseminated in the English-speaking world by Talcott Parsons, a leading American sociologist of the 1950s and 1960s, dissociates 'modernity' from its modern European origins and stylizes it into a spatio-temporally neutral model for processes of social development in general. It thus loses any connection to its origins as a critique of a specific social trend. In this rendering, modernity, thoroughly confused by now with American society, becomes a social model to which other, 'less developed' societies can aspire.

This way of thinking was possible only because time was thought of as coextensive with the world. This took the simple form of here (the West) was now and there (the Rest) was then. All the world was united in a single, one-way history. If technological developments such as the railway and the telephone worked to create a sense of global simultaneity, the understanding of different places in relation to the single trajectory of history worked in the opposite direction, towards an absence of 'coevalness': the sense of societies affecting one another at one and the same time. The idea of temporal evolution determining geographical differences won out over the idea of such differences resulting from the contemporary interplay between local societies and influences emanating from over the horizon.

A final boost to the designation of areas as backward or modern came from the ideological combat of the Cold War in which the two modern worlds of capitalism and communism struggled for dominion in the backward or traditional Third World. Although later adopted as a symbolic referent for the solidarity of former colonized peoples, the term 'Third World' was never a particularly useful empirical designation. Its meaning was premised upon the prior existence of two competing models of development that would allow for no alternatives. This accounts for the origin of this vocabulary in the early 1950s, at the outset of the Cold War between the United States and the former Soviet Union. The backwardness of the Third World was necessary to define the modernity of the other two. In that zone the two modern worlds could compete for converts to their respective models of political economy. In this translation of time into space, the Third World represented what the other two worlds used to be like. Their claims to modernity rested on comparisons with those parts of the world that still had not reached their levels. There could be a modern, therefore, only if there were a backward against which to compare it. Both sides in the Cold War saw their competition with one another in this way. Their pasts were mirrored in the present and future of the rest of the world.

Conclusion: from metaphor to myth

This criticism is all very well, one might say, but surely the vocabulary of backward and modern is nothing more than an evocative metaphor that helps to communicate the differences between Australia or Italy and two ideal types of society – the backward and the modern? I would argue that the vocabulary is now much more than this – that it organizes and directs thinking about the nature of such places as Australia and Italy. From this perspective, primitive Australia and backward Italy are myths, the idealization involved forgotten as the metaphor has substituted for analysis. This is not to say that the metaphor is false in all its usages, say, as a description of differences in technological prowess, only that it functions in the cases examined here, and in the modern geopolitical imagination, as a fable more than as a mere communicative device.

The literary scholar Frank Kermode makes a distinction between 'myth' and 'fiction' that may be helpful here. In his view, a fiction is 'a symbolic construct ironically aware of its own fictionality, whereas myths have mistaken their symbolic worlds for literal ones and so come to naturalize their own status' (Eagleton 1991, 191). The line between the two is fuzzy rather than hard and fast, as all fictions can become myths once established and widely disseminated. This is what has happened to the metaphors of primitive Australia and backward Italy.

It is something of a conventional wisdom among European and American intellectuals that modernized societies (Europe, particularly England, and the United States are the paradigm cases on a global scale) are rational and secular to the exclusion of traditional or metaphysical myths about their founding and nature. As argued earlier in the chapter, this position has become a central element in the backward–modern metaphor itself. Modernity is, by definition, life without myth. But perhaps it is more that our most cherished European myths, such as backward versus modern, are simply ones without hope of providing a suddenly better world in its entirety, merely naturalized fictions that give meaning to speculation about the historical trajectories of particular societies – although it is part of their appeal, at least in the idiom of improvement, that they do hold out the hope that others can become more like us.

It is precisely the *commonsensical* quality of the opposition between the traditional and the modern that gives it mythic power. Since Columbus first returned from his transatlantic voyages we have become so used to characterizing geographical differences in idealized temporal terms that we cannot see any problems with this way of thinking. The explorers of Australia certainly thought this way. They assimilated empirical information to the oppositions (primitive versus civilized, etc.) that they worked with. Later generations have continued to read the whole world through the basic opposition of backward and modern. The message of subordination implied by this pairing is sent masked as one of natural history. So, even as the original context for turning time into space was extra-European, following the early voyages of discovery and conquest, the same message can now be sent inside Europe, indicating that the essential message is still one of geopolitical dominance and subordination rather than justifiable empirical claims about national development. The ultimate irony, therefore, given Columbus's own origins in an Italian city-state (Genoa) and that it is the ancient Romans who largely inspire the elemental distinction between *Homo humanus* (the Roman) at the core and *Homo barbarus* (everyone else) at the periphery, is that in a European context Italy has been more readily portrayed as backward than modern. A temporal metaphor initially applied to make sense of the spatial gap between the new worlds and the old one has become a preferred way of dealing with, for example, Italian differences relative to an idealized European modernity. In so doing, the intrinsically normative character of the terms 'backward' and 'modern' has been obscured. Yet it is their nature as moral judgements about people and places that gives them such power. They tell us who is in charge and why they should be. At least Columbus and his contemporaries seem to have understood that. Perhaps it is time those who operate in terms of the coupling of modern and backward owned up to ethical complacency for ignoring the moral consequences of turning time into space? But that would require abandoning one of the key tenets of the modern geopolitical imagination.

Recommended reading

Chow, R. 1996 We endure, therefore we are: survival, governance, and Zhang Yimou's *To Live*. *South Atlantic Quarterly*, 95: 1039–64.

Fabian, J. 1983 *Time and the Other: How Anthropology Makes Its Object*. New York: Columbia University Press.

Goody, J. 1996 *The East in the West*. Cambridge: Cambridge University Press.

Mandrou, R. 1973 *From Humanism to Science, 1480–1700*. Harmondsworth, UK: Penguin.

Mason, T. 1988 Italy and modernization: a montage. *History Workshop*, 25/26: 127–47.

Pagden, A. 2001 *Peoples and Empires: Europeans and the Rest of the World from Antiquity to the Present*. London: Weidenfeld and Nicolson.

Ryan, S. 1996 *The Cartographic Eye: How Explorers Saw Australia.* Cambridge: Cambridge University Press.

Said, E. W. 1978 *Orientalism: Western Conceptions of the Orient.* New York: Vintage.

Spanos, W. V. 2000 *America's Shadow: An Anatomy of Empire.* Minneapolis: University of Minnesota Press.

Tiley, R. 2001 *Australian Navigators: Picking Up Shells and Catching Butterflies in an Age of Revolution.* Sydney: Kangaroo Press.

4 A world of territorial states

In the modern geopolitical imagination, power has been defined as the ability to make others do something you desire and, at least from the nineteenth century onwards, it has been exclusively associated with territorial states that are usually presumed to be nation-states (ones where there is a close match between membership of a distinctive nation and the boundaries of a particular state). In this chapter I want to explore these contentions in some detail and to show how the spatiality or geographical organization of power is not necessarily tied for all time and all places to the territoriality of states. The state-centred account of spatiality of power, what I term the 'territorial trap', is the historical projection of a world in which power over others is envisioned as pooled up in the hands of equivalent units of territorial sovereignty, usually the most important states militarily, the so-called Great Powers.

Most explicit in the case of political geography and the study of international relations, but common throughout the contemporary social sciences, the conventional understanding of the geography of power is underpinned by three geographical assumptions: first, that states have an exclusive power within their territories as represented by the concept of sovereignty; second, that 'domestic' and 'foreign' affairs are essentially separate realms in which different rules obtain; and finally, that the boundaries of the state define the boundaries of society such that the latter is totally contained by the former. These assumptions reinforce one another to produce a state-centred view of power in which the space occupied by states is seen as fixed, as if for all time. Thinking about the spatiality of power is thus put beyond history by assuming an essential state-territoriality to the workings of power.

This perspective did work particularly well for the state-centred world that began to develop in Europe in the nineteenth century (see the section on naturalized geopolitics in Chapter 6, pp. 93–101). It made sense in the context of that time to see trajectories of economic and social change as increasingly characterized in terms of the experiences of the bits of space delimited by the geographical boundaries of states. Businesses and trade unions, representative politics, and social life were increasingly organized on a state-by-state basis. But there was also a normative element to thinking about power largely in terms of states. As a reflection of the burgeoning nationalism in Europe (and increasingly elsewhere) in the nineteenth century, politics was seen as best thought of in terms of national states. Ideas of distinctive national characters and their reflection in military, sporting, technological, artistic and educational prowess became widely accepted.

This process was long seen as largely the prerogative of Europe (and areas of European settlement). Europe was viewed as having the necessary prerequisites for statehood, particularly such features as an inheritance from the European ancients, the Greeks and Romans; racial superiority; and obvious technological prowess. The rest of the world was available for European imperialism. Between 1800 and 1914 the major European Powers, Russia and the United States went from controlling 35 per cent to controlling 84 per cent of the Earth's surface (Kennedy 1987, 148–49). The world's empires all maintained control by means of acquiescence as much as by force, often operating through local rulers of one sort or another. The example of the United States and the independent republics of Latin America suggested that imperial rule was not destined to last for ever. So, in the end,

> What had applied to the Roman empire also applied to the British, French, German and, ultimately, Russian empires: subject peoples were only willing to remain in subjection so long as at least a significant number of them could see some benefit in doing so.
>
> (Pagden 2001, 163)

Once resistance to colonial rule began, the main means of organizing it was, ironically, 'the same refashioning of society that had been the driving force behind most modern imperialism, namely nationalism' (ibid., 164). Modern statehood, and the modern geopolitical imagination, thus became one of Europe's most popular exports.

At the same time that European nationalism and imperialism were creating circumstances propitious for the spread of statehood around the world, the new social sciences (economics, sociology, political science) used the territories of modern statehood to serve as a fixed and reliable template for their investigations into a wide range of phenomena. The modern (European) world was seen as one in which local communities were in eclipse to the rising sun of societies based on the nation-state. *Gesellschaft* (society) was replacing *Gemeinschaft* (community) as the dominant cultural–geographical ordering principle. In this way a largely implicit 'methodological nationalism' (a phrase first coined by the sociologist Herminio Martins) came to prevail in political and social thought. Currents of thought allowing for more complex views of the geographical scales at which social, economic and political processes could take place were effectively marginalized.

The modern territorial state was underpinned, therefore, by the claim that it was the people's 'mentor' in the cult of the nation. At its extreme, for the French Revolutionary disciples of Rousseau, the nation-state provided the basis for re-establishing a religious foundation to political authority. Only now, unlike in the person of the emperor-god or divinely appointed monarch, the educating state would give citizens a feeling of moral unity and identification with the fatherland (or motherland). This sacralization of the nation gave the territorial state an increasingly competitive advantage over other possible types of spatial–political organization such as confederations, loose empires or city-states.

The drawbacks of a state-centred perspective on power have only recently become obvious. On the one hand, this realization owes much to perceived changes in the ways in which states relate to one another and to the emergence of a global society in which states must share power with other types of actors. We live in an epoch in which the declining military viability of even the largest states, growing global markets, expanding transnational capitalism, and modes of governance alternative to that of the territorial state (such as the European Union, the various UN Agencies, the World Bank and the

International Monetary Fund) have begun to undermine the possibility of seeing power as solely a spatial monopoly exercised by states.

On the other hand, the problem is more profound than one of a mere goodness-of-fit to the changing economic and social conditions of the contemporary world. State-centricity has finally been recognized as the main strategy of modern intellectuals of all political persuasions in limiting the definition of power to that of a coercive instrument and restricting politics to the domestic realm of the state (a way of thinking that has its roots in the works of the ancient Greek philosopher Aristotle as well as those of the early modern Florentine political theorist Machiavelli). Representing space as state territoriality also serves to put statehood outside time, because of the strong tendency to associate space with stasis or changelessness, and thus to impose an intellectual stability on the world that would otherwise be difficult. As a result, state-centricity has continuing normative attractions for both intellectuals and political activists even as the empirical reach of states to control and regulate recedes. It provides a grounded set of social-geographical units for both longitudinal and cross-sectional data analyses. It offers a set of concrete institutional opportunities (however weakened or compromised in effective performance) for political action.

The territorial trap

Three analytically distinct but invariably related assumptions underpin the territorial trap: thinking and acting as if the world were made up entirely of states exercising power over blocks of space that between them exhaust the politico-geographical form of world politics. The first, and most deeply rooted, is that modern state sovereignty requires clearly bounded territorial spaces. The modern state differs from all other types of organization by its claim to total sovereignty over its territory. Defending the security of its particular spatial sovereignty and the political life associated with it is the primary goal of the territorial state. Vested at one time in the person of the monarch, or other leader within a hierarchy of orders from the lowest peasant to the warriors, priests and nobles, sovereignty is now vested in territory.

The second crucial assumption is that there is a fundamental opposition between domestic and foreign affairs in the modern world. This rests on the view common to Western political theory that states are akin to individual persons struggling for wealth and power in a hostile world. One state's economic and political gains always come at the expense of others. Only inside the boundaries of the state, therefore, are civic culture and political debate possible. Outside, reasons of state (pursuit of the state's interests) rule supreme. This view fixes processes of political and economic competition at the level of the system of states.

Third, and finally, the territorial state acts as the geographical container of modern society. Social and political organization are defined in terms of this or that state. Thus, we speak and write unselfconsciously of 'American' or 'Italian' society, as if the boundaries of the state are also the boundaries of whatever social or political process we might be interested in. Other geographical scales of thinking or analysis are thereby precluded. Often this is because the state is seen as the guarantor of social order in modern societies. The state substitutes for the self-reproducing cultural order that can be found in so-called traditional societies (this ties back into the theme of Chapter 3).

Together these three assumptions underpin a timeless conception of statehood as the unique font of power in the modern world. The first dates from the period in European history when sovereignty shifted from the person of the monarch to the state and its citizens. In Europe, this did not happen overnight; it was a process lasting from the fifteenth to the nineteenth centuries. The second two date from the past hundred years, although the domestic versus foreign opposition has roots in the doctrines of seventeenth-century economic mercantilism. Together they serve to put the modern territorial state beyond history in general and the history of specific states in particular. They define a world made up exclusively of similar territorial actors achieving their goals through control over blocks of space.

A case in point: standard international relations theories

The importance of the territorial state and the similar roles it performs within different theories can be seen in the writings of such influential contemporary international theorists as Kenneth Waltz (1979) and Robert Keohane (1984). They are influential exponents at either end of the continuum stretching from realism to liberalism – the main positions on state power around which the modern geopolitical imagination tended to operate in the twentieth century. Waltz is concerned with what he calls the 'structures of inter-state relations', altogether excluding the domestic character of states from explanatory consideration. To him, the structure of the international system has three features that count: it is anarchic, without higher authority; states all perform the same functions and are equivalent units; and there is an uneven distribution of resources and capacities among states. From these fundamental features he draws the following inferences: that at any moment the system's shape as a whole is determined by the number and relations between the Great Powers (the ones with most resources and capacities) and that the balance of power between the Great Powers is the key mechanism in world politics. From this point of view, therefore, from 1945 until 1990 the international system was a bipolar balance of power between two Great Powers (the United States and the Soviet Union) in contrast to, for example, the multipolarity of the early nineteenth century in Europe. What drives the system is fear of domination by others. States are thus understood as unitary actors with each state trying to maximize its status relative to that of others (see Chapter 5 for a longer discussion of how important this has been for the modern geopolitical imagination). No entities other than states are involved, by definition, in international relations. World politics is entirely about international (i.e. inter-state) relations.

In apparent contrast, Keohane is interested in how cooperation can take place between states without there being a dominant Great Power. He argues that there are important incentives for cooperation between states that work against the competitive pursuit of coercive power in an anarchic world, even though he accepts the reality of this world. Treaties, agreements and formal international institutions do restrict state conduct. This is because states agree to constraints when the benefits that derive therefrom outweigh the costs. States are seen as utility rather than status maximizers. But states remain the only significant actors in this account. The utility game is one only they play.

Despite their differences over status or utility maximizing, therefore, these theorists share the commitment to a state-centred world or, more precisely, a Great Power-centred world. This is a conception inherited from a long line of political thinkers and practitioners. It is a vital ingredient of the modern geopolitical imagination.

The spatiality of power in the modern geopolitical imagination and its limits

Coercive power over blocks of space

Three features of the conventional view of power and states have been crucial in limiting understanding of the spatiality or geographical organization of power entirely to states. These made the territorial or spatial block perspective on power problematic long before recent changes in the workings of the international political economy called it more fully into question. The first feature has been the definition of power implicit in the modern geopolitical imagination as the capacity to coerce others within your territory into doing your will (or power *over*). This leads to a notion of power as a monopoly of control exercised equally over all places within a given territory or geographical area by a dominant social group or élite (*despotic power*). Missing from this account are the contingency and fragility of the *infrastructural power* (state provision of public goods and services, etc.) upon which the legitimacy (the right to rule recognized by a population) of modern states largely rests. The sociologist Michael Mann (1984) has pointed out in some detail the critical role of infrastructural power in distinguishing modern territorial states (both bureaucratic and authoritarian) from both feudal and classic imperial types of rule (Table 4.1). With the ability to provide centrally and territorially organized services that other organizations cannot, the territorial state is no longer entirely the creature of state élites (and their despotic power). It has an autonomous source of power in its coordinating and directing roles. But this relative autonomy depends upon states delivering a set of services that cannot be provided in some other way. This, of course, opens up to challenge both the regime (current institutions) and the state if the state cannot be depended on to deliver the goods.

The redefinition of power in terms of infrastructural and despotic power, however helpful in pointing to the dependence of modern state power on what it does for its population, completely ignores the degree to which power is inherent in all human agency. All social practices involve the application of power – the ability to engage in actions towards the completion of socially sanctioned goals (or power *to*). From this point of view, power is not some thing or potential vested solely in states (or associated political institutions) but the application of agency inherent in all social action to achieve chosen ends. Territorial states are one type of concentration of social power that emerged in specific historical conditions in which state territoriality was practically useful in fulfilling the objectives of both dominant and subordinated social groups. Today, we see the emergence of local and regional governments and supra-regional

Table 4.1 *Two dimensions of state power and the four ideal types of state they define*

		Infrastructural power	
		Low	*High*
	Low	feudal	bureaucratic
Despotic power			
	High	imperial	authoritarian

Source: Adapted from Mann (1984, 188)

communities in the application of infrastructural power for such ends as economic development and political identity, usually without the coercive power traditionally associated with territorial states. Of course, such alternative spatial configurations of power to that of the modern territorial state are not entirely new. For example, the Hanseatic League, the Swiss Confederation, the Holy Roman Empire, the Iroquois Confederation, the Concert of Europe and the early United States are familiar examples of systems of power and authority alternative to that of the Westphalian system of territorial states. The actual existence of these kinds of institutional arrangements points to the range of possible ways in which power can be organized spatially. They suggest that forms of power are generated, sustained and reproduced by historically and geographically specific social practices, rather than given for all time in one mode of spatial organization: that of state territoriality. Indeed, there was nothing inevitable about the emergence of the modern system of territorial states. Until the nineteenth century even their monopoly over coercive power was easily challenged, for example by pirates, and alternative arrangements for the geographical organization of centralized power were widespread, for example in overlapping jurisdictions such as the Holy Roman Empire.

More radically, however, the power of states over their populations and in relation to one another can be understood as resting largely on power 'from below'. In other words, the territorial state draws its power in capillary fashion from social groups and institutions rather than simply imposing itself upon them. From this point of view, power is present in all relationships among people, animals and things, and the power of the state relies on the wide range of sources it can tap into. This can be termed a non-sovereign conception of power, in contradistinction to that view which sees power as flowing from a single (sovereign) source, such as the state. In this construction, power is best thought of as equivalent to the energy moving a circulatory system rather than a mechanical opposition between a source of power, on the one hand, and an obedient (or truculent) subject, on the other. There are multiple points at which consent and resistance come into play in expanding and restricting the interplay between states and subjects, and hence in defining the state's effective territoriality: how well it dominates its claimed block of space. The spatial monopoly of power exercised by a state is not and cannot be total when its power derives from that given up by and potentially retaken by others.

Both the importance and the limits of conventional accounts focusing on coercive power exercised over a state's territory can be seen with reference to the 'nation-states' of Africa. In the mid-1960s, just as many of the former European colonies in Africa had come to independence, a 'breakthrough' or escalation in sovereignty was widely heralded. Thirty-five states acquired the trappings of state sovereignty between 1951 and 1966. But this turned out to be illusory for two reasons. First, whatever the doubts of individual nationalists, imperialist policies obliged them, as a practical necessity, to seek independence within existing colonial territories that had neither the cultural building materials for nation formation nor political boundaries that made either economic or geopolitical sense. The end result has been a pervasive politics of clientelism as state managers facing deteriorating terms of trade, massive rural–urban migration, and the failure of initiatives to develop home-grown industries and a middle class have milked state coffers (and the proceeds of exporter groups) to legitimize their rule, at least in the eyes of their beneficiaries. The import of Cold War ideologies in grotesque forms attracted foreign aid and advisers but proved of little long-term benefit. The ability of states to police or to develop their territories – or, in other words, consistently to exercise either despotic or infrastructural power over them – has remained beyond their grasp. The net effect has been to make the state a barrier to, more than an agent of, progressive social change in large parts of Africa. Second, since independence, African states have not been regarded by the world's Great Powers as the equivalents of European or other 'real' states with respect to sovereignty. So it is not simply the absence of internal efficacy but also the external lack of respect that undermines more effective African statehood. The apparent social 'chaos' of Africa is seen as calling for external intervention or management once more by Europeans. It is as if there are distinct 'regimes' of sovereignty that prevail in different parts of the world. From this perspective, Africa's fate was set during the Cold War. Not only did the continent become the site of much Cold War gamesmanship with episodic interventions and surrogate power struggles, but its recent history of colonialism – with continuing conflicts between natives and settlers in Zimbabwe, South Africa and elsewhere down into the 1990s – also marked it off as a zone in which statehood was seen as a sort of Potemkin village: a cardboard imitation of the real thing. This view – of Africa as a 'wild zone' beyond the 'normal' politics of Westphalian statehood – is now widely shared by both many Africans and outsiders.

Power as coercion in international relations

The second feature of the association between states and power in the modern geopolitical imagination is that coercive relations between states have usually been seen as the only way in which power is exercised beyond state-territorial boundaries. Even cooperation between states, as in the Keohane example cited earlier, is interpreted as 'getting your own way' in disguise! Indeed, the practices of politics, group divisions and struggles over 'the good society' and who gets what, when, how and where, within state boundaries, are typically contrasted with reasons of state (expedient action to defend or further the interests of the state itself) beyond them. Democratic political theory, for example, has been largely limited to the possibilities of political representation and participation within states and not their prospects among, between or beyond them. At the level of the state system, the concept of *hegemony* is often used to indicate the domination exercised by a particular state over others during a specific historical epoch. This definition is indicative of the central position given to despotic or coercive power in international relations, irrespective of the specific 'relation' in question.

Yet the concept of hegemony can be given a different meaning, closer to one originally suggested by the Italian Marxist thinker Antonio Gramsci (1891–1937), that refers to the power implicit in dominant practices that govern society, both within and beyond state-territorial boundaries. In this construction, therefore, world politics involves a variety of social practices that require the deployment of power, not simply military coercion by states. The identities and interests of states (and other actors) are formed in interaction with one another and in the nexus between global and local social practices. Hegemony refers to the nature of the dominant social practices in a given historical epoch and how they bind together the various actors into a global society. Power can involve gaining assent, defining expectations, and co-opting others as much as or more than simply coercing them. It is not a singular entity but can involve so-called *soft* forms of consent as well as *hard* forms of coercion. The dominant practices may benefit one state disproportionately (such as Britain in the mid-nineteenth century and the United States since the Second World War; see Chapter 6), but the costs and benefits (both economic and cultural) can be more diffusely distributed among all the actors (both state affiliated and otherwise) subscribing to the contemporary 'principles' of international life – such as those defining the modern geopolitical imagination – irrespective of their geographical location. This is one of the insights emanating from so-called post-colonial studies, drawing attention to the worldwide penetration of dominant practices and understandings (such as those labelled 'nationalism') and their naturalization into the routines of everyday life as 'common sense' and 'facts of life'. In a world of social practices rather than reified institutional forms, therefore, not only must states play by 'rules' established by dominant social groups that are active in all of them, but the nature of rule making presupposes that states (and other actors) are not simply coercive agents in a world of anarchy.

Two examples serve to illustrate why this argument matters. The first concerns the changing social and technological conditions for the military viability of states and their impact on the possibility of war serving reasons-of-state. The advent of nuclear weapons has meant that security now 'derives from the paralysis of states rather than from the exercise of state power, and from the acceptance of the impossibility of territorial violence monopolization rather than its pursuit' (Deudney 1995, 219). At the same time, the spread of easy-to-use conventional weapons (Kalashnikov machine guns and Stinger surface-to-air missiles, for example) has made it much easier for local populations to resist the designs of apparently more powerful adversaries. As Deudney puts it in summarizing the point about contemporary military practices, 'It is nearly impossible to protect territory from annihilation; but it is easier than ever to prevent conquest' (ibid.).

The second involves the contemporary decentring and deterritorialization (at a state level) of the means of production and communication. This reflects the opening up of the world economy for increased cross-border flows of trade and investment under US auspices during the Cold War. Recent developments in financial markets and information technologies, however, have accelerated changes in the ways people, places and states interact and how economic and political actors perceive these interactions. States (and others) must now manage these interactions. Though external coercion is a real possibility for the most powerful states, it is now of limited use when state policies must be concerned with attracting foreign capital and gaining access to global flows of information. Contemporary economic practices, therefore, point towards the incipient creation of a transnational–liberal hegemony (see pp. 61–64) in which territorial states are no longer the basic building blocks; they are being rapidly challenged by new spaces of networks and flows in which speed and access are more important than command over territory. Even in France, for example, often pointed to as a zone of changelessness by those sceptical of claims about an emerging world of flows challenging that of territories, recent neo-Gaullist and Socialist governments supported a new European currency and a withdrawal of that state from many areas of life. It was General de Gaulle who said that France exists only because of the state, the army and the franc. His disciples and his political enemies now both stand watch over their dismantling.

Statehood and the protection of property rights?

When states are situated in the context of a world of evolving social practices, they lose their exclusivity. But one historic role of states is thereby re-emphasized. This is that of the definition and regulation of property rights. The modern territorial state system has been associated from its origins in Europe in the seventeenth and eighteenth centuries with the framework for definitions of property rights (legal rights of ownership and use), without which global capitalism would not have been possible. States are never so sovereign, in the conventional sense of singular entities endowed with power monopolies within their territories, as when they are seen as definers and enforcers of property rights.

The third feature of state-centred accounts of the spatiality of power, therefore, is that they are silent as to the role that states have played in the growth of certain basic social practices of capitalism – defining and protecting private property rights – that have inexorably led beyond state boundaries in pursuit of wealth from the deployment of 'mobile property' (capital). The term 'property' implies a fixity or permanence in place that modern territorial states have given a high priority to protecting. Consequently, much of the law in most Western states is devoted to establishing rights of ownership and access. But a home territory also provides a base from which to launch attempts at acquiring property elsewhere. This requires that assets be reasonably liquid and transferable over space and across state boundaries. At a certain point, however, states endure a tension, what Ruggie (1993, 164) calls the problem of 'absolute individuation', which can give rise to an 'unbundling' of territoriality when states effectively exchange control over economic flows emanating from their territories for increased access to flows coming from elsewhere. As a result, when increasing proportions of property are mobile beyond any one state's boundaries, individual states provide only a partial and tenuous protection for absolute property rights. Other geographical levels of governance and regulation become attractive, as was the case with the Bretton Woods system that regulated international finance from 1944 to 1972 and is currently the case (if signally less effectively) with the annual G-8 summits between the leaders of the Big Seven industrial countries. But uncertainty as to future political actions and macroeconomic changes (tariffs, interest rates, etc.) also gives an incentive to property-holders to further spread assets around rather than leave them pooled up in one state.

This process is not new. Its origins go back to the merchant capitalism of the sixteenth century. What is new is the increased quantitative scale and the enlarged geographical scope of the mobile property now moving to and fro across the boundaries of the world's trading and investing states. In this context, states and firms have changed their orientation from free trade to what has been called 'market access' (Cowhey and Aronson 1993). The underpinnings of the world trade regime that prevailed in the aftermath of the Second World War are being replaced by those of a regime in which a premium is placed on the openness of borders. 'Leakiness' in cross-border flows of goods and investment and in firm multinationality has become a torrent of capital, trade and corporate alliances. Cowhey and Aronson (ibid., 237) contrast the nature of the old regime with that of the new one by identifying the six pillars that they claim have underpinned each and the policy instruments associated with the new regime (Table 4.2). The policy instruments reflect an abandonment of classical state sovereignty in return for guaranteed rights of access to other states' territories. The world has moved away from the strict association of property rights and capital accumulation with state territoriality. A range of non-territorial factors now determine the competitiveness of firms in many industries: access to technology, marketing strategies, responsiveness to consumers, flexible management techniques. All these are now the assets of firms, not territories. Firms grow through deploying their internal assets as successfully as possible. States now compete with one another to attract these mobile assets (property) to their territories.

Three aspects of the market-access regime are particularly notable with respect to the changing spatiality of power. One is the internationalization of a range of domestic policies to conform to global norms of performance. Thus, not only trade policy but also industrial, product liability and social welfare policies are subject to definition and oversight in terms of their impacts on market access between countries. Another is the increased trade in services, once produced and consumed largely within state boundaries. Partly this reflects the fact that many manufactured goods now contain a large share of service inputs – from R&D to marketing and advertising. But it is also because the revolution in telecommunications means that many services, from banking to design to packaging, can now be provided to global markets. This represents a significant material challenge to the domestic versus international distinction upon which the 'realism' of strictly territorial accounts of the spatiality of power relies. Finally, the spreading reach of transnational firms and the emergence of international corporate alliances have had profound influences on the nature of trade and investment flows, undermining the identity between territory and economy. Symptomatic of the integration of trade and investment are such frequently heard concerns as rules on international investment and unitary taxation, rules governing local content and place of origin to assess where value was added in the commodity chains of globalized production, and rules involving unfair competition and anti-trust or monopoly trading practices.

Table 4.2 *Pillars of the emerging 'market-access' world economy*

Pillars of the free trade regime instruments	Pillars of the market-access regime	Policy
Governance		
1 US model of industrial organization	Hybrid model of industrial organization	More reliance on bilateral and plurilateral negotiating forums
2 Separate systems of governance	Internationalization of domestic policies	Transparency, specialized rights of appeal, and self-binding behaviour
3 Goods traded and services produced and consumed domestically	Globalization of services; eroding boundaries between goods and services	National treatment and tiered reciprocity for services
4 Universal rules are the norm	Sector-specific codes are common	Reforms of voluntary restraint agreements, anti-dumping and subsidy codes
Rules		
5 Free movement of goods; investment conditional	Investment as integrated co-equal with trade	Rules of origin and new investment rules to ensure market access, global anti-trust
6 National comparative advantage	Regional and global advantage	Fair trade rules for procurement, standards and R&D

Source: Adapted from Cowhey and Aronson (1993, 237)

Transnational liberalism and new spatialities of power

Outlining the three main features of conventional accounts of the spatiality or geographical organization of power brings into focus their common deficiency: taking the coercive power of territorial states for granted as a fixed feature of the modern world rather than seeing it as the outcome of a number of historical contingencies. The contemporary unbundling of state territoriality provides the most direct evidence for the reshaping of hegemony away from the state-centred practices of the previous epoch. This does not mean to say that territorial states are (finally) withering away, only that they must now operate in a global context in which their interactions with one another must now take into account a changed military and economic environment. Indeed, in the absence of higher-level units for the enforcement of property rights and the delivery of public goods, states have a continuing and vital role to perform within the evolving world of networks and flows. For example, the deregulation of financial markets requires the deliberate action of governmental authorities. It does not simply 'happen'. During the Cold War between 1947 and 1990, the United States, in competing militarily and ideologically with the Soviet Union, sponsored an unprecedented opening of the world economy, partly to spread its political–economic 'message' and partly to take advantage of opportunities for its businesses. The net effect has been that markets have acquired powers heretofore vested in leading states. As this process has intensified and expanded, some localities and regions within states have been privileged within global networks of finance, manufacturing and cultural production to the disadvantage of others. The 'market-access regime' ties local areas directly into global markets. Successful ones are those that can enhance their position by increasing their attractiveness to multinational and global firms. A patchwork of places within a global node and network system therefore coexists with but is slowly eroding the territorial spatiality with which we are all so familiar. Two consequences of this trend are illustrated in the rest of this chapter to give some empirical substance to the claim that a new geopolitics of power is in the offing: the increasingly fraught intersection of migration and citizenship, and the increasingly decentralized world financial system, illustrated by the deterritorialization of currencies.

Migration and citizenship

Modern citizenship is closely connected to the rise of the state. Conventional views of statehood see control over membership in its territory as a crucial requirement. Exclusive loyalty to a specific state is likewise seen as a sine qua non of citizenship. Statehood has been conjoined to nationhood by means of citizenship. Political participation is also an essential component of democratic citizenship. Struggles to extend and deepen political representation focused on democratic control of state institutions have served to give states one of their most important sources of legitimation. Consequently, most democratic theory and practice assumes a territorial political community with citizenship as a means of delineating who does and who does not belong to the 'people'. Today, only states have the authority under international law to grant or deny the status of citizen. Thus, citizenship is strongly linked to the idea of political community, which in turn is seen as synonymous with the territorial exclusivity of the sovereign nation-state.

But the historic tie between states and citizenship is under increasing pressure from immigration and the development of multinational and global conventions governing human rights (such as those established by the European Union). First of all, rights of residence as opposed to birthright are increasingly driving definitions of citizenship. In Europe and North America a case can be made for a 'paradigm (and scale) shift' in understandings of citizenship, a shift that relocates citizenship from nation-state sovereignty to the international human rights regime. This reflects both pressure from the absolute numbers of immigrants and the fears of labour unions and other interests that without some rights of political membership for immigrants the rights of all will be undermined. Yet at the same time, political rights that extend across borders from one state to others are also growing. Citizens of a country resident abroad now can have voting and pension rights hitherto restricted to those resident within the country in question. This is now the case, for example, with Mexican citizens resident in the United States. Within the European Union the issue of dual or even multiple citizenship has been transcended by the possibility of a European citizenship that allows for continuing allegiances at not only the state but also at sub-national scales of identity. Plural citizenship, therefore, is an emerging reality.

Recent international migration is distinctive from that in the past in two ways that are particularly threatening to traditional conceptions of citizenship. One is in the long-term concentration of migrant communities in certain cities and localities that, rather than assimilating into a national 'mainstream', maintain their cultural particularity. This is a result of both greater cultural pluralism and tolerance in host countries and greater cultural differences between the new immigrants and their host societies. Many immigrants remain attached to their 'homelands' and see themselves as temporary absentees more than permanent migrants. Another distinctive feature of contemporary global migration is the ease of movement of people and ideas between source and destination areas. With the new telecommunications technologies, it is relatively easy today to keep ties across state boundaries and to develop political and economic attachments without a 'final commitment' to one state or the other.

As definitions of citizenship are disrupted by the novel character of contemporary global migration, there are countervailing pressures to re-establish 'normalcy'. 'Invasion panics' based on exaggerated fears about the scope and impact of immigration have afflicted such disparate destinations as California, France and Italy over the past ten years. Often these are cultural in inspiration, given the increase in flows of migrants from poorer, underdeveloped countries to richer, developed ones who are more visibly different from native populations than previous generations of immigrants. But they also reflect economic concerns about job competition or burdens on public-sector spending for welfare or social security. Political parties play the 'immigrant card' in certain areas and constituencies when they use the 'threat' of immigration to mobilize native voters. At a certain point in time, however, this strategy can backfire, as it has for the Republican Party in California, when sufficient numbers of immigrants have acquired citizenship and demonstrate their electoral strength by voting en masse against those who would demonize immigrants as the dominant source of local social and fiscal woes.

Citizenship is a core feature of state sovereignty. Whether democratic or not, states rely on a high degree of exclusivity of identity drawn by their citizens to maintain power within their jurisdictions. Historically, some civil and social rights have been granted to non-citizens. Increasingly, however, even political rights have become relatively mobile. Non-resident citizens, immigrant citizens, residents of encompassing jurisdictions, such as the European Union, and multiple citizens are categories of people who experience citizenship in ways that violate the one-to-one correspondence of state and citizenship upon which state licensing of political power has long rested. One of the great advantages of states, to speak and act on behalf of nations, is undermined when the key link between the two, an affective and singular citizenship, is eroded by movements of people that transgress rather than reinforce the boundaries of states.

Money and states

The control and maintenance of a territorially uniform and exclusive currency are often regarded as one of the main attributes of state sovereignty. If a state cannot issue and control its own currency then it is not much of a state. Benjamin J. Cohen (1977, 3) offers a concise statement of this view:

> the creation of money is widely acknowledged as one of the fundamental attributes of political sovereignty. Virtually every state issues its own currency; within national frontiers, no currency but the local currency is generally accepted to serve the three traditional functions of money – medium of exchange, unit of account, and store of value.

Currency has a further and vital symbolic role in underwriting statehood:

> As Keynes [the famous economist] understood, the creditworthiness of a nation's money is perhaps the primary evidence to the faithful (the citizens) that the ultimate object of their faith, the nation-state, is real, powerful and legitimate; it is the ultimate 'guarantor of value'.
>
> (Brantlinger 1996, 241)

Yet over the past thirty years a number of trends have challenged the notion that every state must have its own territorial currency (currencies that are homogeneous and exclusive within the boundaries of a given state). This does not necessarily portend a crisis for the state system as such so much as a challenge to that legitimacy of states which rests on claims to represent particular nations and associated national interests by means of control over singular currencies. Territorial currencies developed on a large scale only in the nineteenth century, once the Westphalian system of states was already in place (Helleiner 1999). Symbolically, however, currencies (including the symbols found on coinage and banknotes) were important elements in establishing central state legitimacy long before the nineteenth century. Modern statehood was not achieved independently of processes of nation building, even though 'state' and 'nation' can be distinguished analytically, the former referring to a set of institutions ruling over a discrete territory and the latter signifying a group of people who share a sense of common destiny and occupy a common space. The fact that the construction of territorial currencies was largely a nineteenth-century phenomenon, therefore, should not detract from the persisting linkage over many centuries between currency and statehood, however ineffective in practice that linkage often was.

Three monetary developments have begun to delink currencies from states in the way they were once largely mutually defining. The first is the growing use of foreign currencies for a range of transactions within national currency territories. The best known of these is the growth of the so-called eurodollar markets in London and other European financial centres. Others would include the development of offshore financial centres such as the Bahamas and the Cayman Islands largely devoted to exchanging, sheltering and laundering foreign currencies. This trend is one part of that set of processes leading to global financial integration, at least among the world's richest economies.

The second is the emergence of projects such as that in the European Union to restrict or abolish national currencies in favour of supra-national or world-regional currencies. In practice, the US dollar, the European Currency Unit (now the euro), the German mark (until it was replaced by the euro) and the Japanese yen have served as transnational currencies for some years. Much of world trade is denominated in one or other of these currencies, irrespective of its particular origins or destinations. Currencies such as the US dollar and the French franc also came to dominate large regions beyond their borders, the dollar in Latin America and the French franc in parts of West Africa that formerly were part of the French Empire. Some of this is the result of internationally mandated economic reform, while some is more the result of local élites keeping their funds in harder (more stable and reliable) currencies. The arrival of the euro in 2002, eliminating most of the existing European territorial currencies, suggests that the process of transnationalization of currencies will intensify in years to come.

Third, and finally, a number of uses have appeared in recent years for 'local currencies', forms of scrip and token money, that substitute for regular national currency. Such uses are often the result of experiments in local communities (e.g. Ithaca, NY, and Montpelier, VT, in the United States) and consumer cooperatives or tokens issued by firms for their products or services. As yet, they cannot really be seen as posing a major threat to territorial currencies. But they are often symptomatic of the lack of trust that territorial currencies now elicit in some quarters, perhaps as a result of the uses of monetary policies that have disadvantaged some groups (and localities) when currencies are rapidly revalued or high inflation persists which pushes people out of the official monetary economy and into a black or underground economy where barter, trusted foreign currencies (such as the former German mark in many parts of Eastern Europe) or local currencies prevail.

The deterritorialization of currencies, therefore, has three aspects to it: the explosion of foreign currency transactions within the territories of hitherto territorial currencies; the rapid increase in the number of economic transactions involving supra-national currencies; and the growing use of local currencies. None of these should be read as totally undermining existing territorial currencies. The continued erosion of territorial currencies will take place only if states continue to allow it. That there is still advantage in it for the most 'powerful' currencies means, however, that it will probably continue. As it deepens it could well gain a momentum that even the most powerful of states will find difficult to counteract.

Conclusion

At one time it made sense to some to see the path of history or social change as a series of 'stages' (as, for example, in Rostow's (1960) famous account of 'the stages of economic growth') inscribed upon state territories. Today, however, economic development and social change are increasingly determined by the relative ability of localities and regions to achieve access to global networks. In this context, understanding power as if it is attached singularly and permanently to state territories makes no sense. But the commitment to an unchanging spatiality of power retains considerable appeal. Not only does it allow for a restriction of politics to an unproblematic domestic space, it also provides an attractive intellectual and political stability by equating space with the fixed territories of modern statehood, which can then serve as a template for the investigation of other phenomena or as the basis for organizing political action. Putting state territoriality in question undermines the methodological nationalism that has lain behind the workings of both mainstream and much radical social science. The major social sciences in the contemporary Western university – economics, sociology and political science – were all founded to provide intellectual services to modern states in, respectively, wealth creation, social control and state management. It is hardly surprising, therefore, that they find difficulty in moving beyond a world unproblematically divided up into discrete units of sovereign space. Political geography would seem to have less of an excuse. Putatively concerned with the spatiality of power, it has remained largely attached to a modern geopolitical imagination that has relied on seeing coercive power and the territorial state as twin markers of modernity. To correct this vision is a major challenge facing political geography at the close of the twentieth century.

Recommended reading

Agnew, J. 1994 The territorial trap: the geographical assumptions of international relations theory. *Review of International Political Economy*, 1: 53–80.

Agnew, J. 2001 *Reinventing Geopolitics: Geographies of Modern Statehood*. Heidelberg, Germany: Hettner Lectures, Institute of Geography, University of Heidelberg.

Allen, J. 2002 *Lost Geographies of Power*. Oxford: Blackwell.

Biersteker, T. and Weber, C. (eds) 1996 *State Sovereignty as Social Construct*. Cambridge: Cambridge University Press.

Cohen, B. J. 1998 *The Geography of Money*. Ithaca, NY: Cornell University Press.

Davidson, B. 1992 *The Black Man's Burden: Africa and the Curse of the Nation-State*. New York: Times Books.

Helleiner, E. 1999 Historicizing territorial currencies: monetary space and the nation-state in North America. *Political Geography*, 18: 309–39.

Jacobson, D. 1996 *Rights across Borders: Immigration and the Decline of Citizenship*. Baltimore: Johns Hopkins University Press.

Money, J. 1997 No vacancy: the political geography of immigration control in advanced industrial countries. *International Organization*, 51: 685–720.

Nye, J. 1990 The changing nature of world power. *Political Science Quarterly*, 105: 177–92.

Ruggie, J. G. 1993 Territoriality and beyond: problematizing modernity in international relations. *International Organization*, 47: 139–74.

Spruyt, H. 1994 *The Sovereign State and Its Competitors: An Analysis of Systems Change*. Princeton, NJ: Princeton University Press.

Teschke, B. 2002 Theorizing the Westphalian system of states: international relations from absolutism to capitalism. *European Journal of International Relations*, 8: 5–48.

Walker, R. B. J. 1993 *Inside/Outside: International Relations as Political Theory*. Cambridge: Cambridge University Press.

5 Pursuing primacy

Within the modern geopolitical imagination there is a logical tension between the normative claim of an essential equality of statehood between all states implicit in state sovereignty and international law, on the one hand, and the historical reality of dramatic
inequality in power between them, on the other. This has been resolved intellectually
and practically by seeing the normative claim of sovereign equality as an initial situation akin to a condition of nature, with inequality and the resulting hierarchy of states
as the result of an inevitable process of competition between states once 'social life' has
begun. After reviewing the classic argument for why a hierarchy of states is a 'normal'
condition in world politics, this chapter identifies the main axioms upon which this position relies, the historical–geographical conditions under which it made sense, and the
difficulties facing it under contemporary political–economic conditions.

The social origins of Great Powers

The position of initial equality giving rise to subsequent inequality is foreshadowed in
Hegel's famous discussion of lordship and bondage, in which nominally equal persons
or selves are seen as locked into an unequal and enduring hierarchy. This is so because
a 'single consciousness' can know itself only through an other, even in a situation of
radically unequal power. So, a lord is a lord only through a relationship of mutual recognition with a bond servant. The character of the relationship, however, is one in which
'the other' is sublated or annulled. Lordship derives from the conquest and negation of
the servant. Recognition of the other is required, therefore, for its negation. A lord must
have a servant who recognizes him to be a lord. In return, the lord must recognize the
selfhood of the servant. Paradoxically, therefore, the mastery or primacy of one over
another implicitly recognizes the separate selfhood of the other, even as the other is
diminished. This is how primacy emerges within a world of nominally equal selves.

If states are substituted for selves, this is the understanding of what drives world politics that has predominated within the modern geopolitical imagination. The paradox
between normative equality of sovereign states and the real inequality between them
has been resolved by presuming an acquisition of primacy among states equivalent to
the status allocation emanating from the lordship/bondage model. Higher-order states or
Great Powers, therefore, emerge as a result of competition between states but can be
accepted as such only if recognized by those states that are subordinated. Primacy
depends in equal parts on successful competition and on subsequent recognition of that
success by other states.

In the politically fragmented Europe in which the modern system of territorial states developed, military prestige was the main measure of competitive success. States formed a status system analogous to that found among social groups. Each sought to emulate the more prestigious and more modern states above them in the hierarchy. Hence, each state sought advancement by competing with other states. Modernization entailed the endless emulation of successful Great Powers by aspiring ones. Imperial expansion has been one strategy by which to claim or maintain the status of being a Great Power. Depending on how you look at it, akin to whether you see a glass as being half-empty or half-full, every state's goal, therefore, is either to achieve primacy or to avoid subordination.

The history of arms races is perhaps the best example of the pursuit of primacy. As each state desires primacy and/or fears subordination, any escalation in the military capacity of a potential adversary engenders a further escalation by way of response. This was the story of the 'battleship rivalry' between, first, Britain and France in the mid-nineteenth century and, second, Britain and Germany from 1894 until 1914. The United States and the Soviet Union followed much the same logic during the Cold War from 1947 until 1991, with each trying to outgun and 'out-missile' the other. The famous American apostle of sea power, Alfred Thayer Mahan (1840–1914), achieved his fame in the late nineteenth and early twentieth centuries largely because of his strident endorsement of the view that the United States had to have a large globe-spanning navy or risk political eclipse or even invasion at the hands of those with larger, long-distance fleets. He based his claim upon an appraisal of an earlier period (1660–1783) but projected this into the indefinite future. His views proved popular with many world leaders, including Kaiser Wilhelm II of Germany, who shared his belief in the strategic value of sea power to achieve worldwide domination.

This view and the practices that relate to the pursuit of primacy depend for their veracity upon a historically fixed notion of statehood and a world in which modernity is in short supply relative to perceptions of backwardness (positions outlined in Chapters 4 and 3, respectively). Accepting the identity of states as self-sufficient actors with clearly bounded 'selves' is the most important move in both representing and practising the pursuit of primacy. From this point of view, states have a 'first person' identity akin to that of individual persons. This is why it has often made sense to talk of 'France' doing this or 'India' doing that, as if each were a capable actor in its own right. One of the most powerful metaphors in modern political theory is the idea of the state as an 'organism' or autonomous entity having a superordinate identity that cannot be reduced to any of its parts (its populations, social groups, etc.). This point of view, with its origins in ancient Greek philosophy, was given an explicitly biological cast in the late nineteenth century by authors such as one of the founders of political geography, Friedrich Ratzel, and reached its zenith with the Nazis. It opens the state to analogical reasoning; to treatment as if it were a person or a biological individual of some type. This organic reasoning became a vital part of the modern geopolitical imagination, particularly as it developed an affinity for naturalized (fact-of-life) explanation in the late nineteenth century.

A second move sees 'social life', once the initial state of nature has passed, as inherently competitive. Though the state provides a means for restricting conflict and encouraging cooperation within its territorial boundaries, beyond those boundaries is in essence a pre-social world in which the English political philosopher Thomas Hobbes's (1588–1679) 'struggle of all against all' goes on unabated. All economic and social welfare is seen as depending on expanding your state's capacity for violence relative to that of other states. This view has ancient European roots. The stress on the naturalness of danger and violence from strangers was used by the influential theologian St Augustine to justify war. He turned the Western Christian tradition away from pacifism by claiming that violence was intrinsic to human nature and could be channelled into righteous paths if it were used to convert the heathen and destroy the heretic. This political theology became widely accepted in Europe and underpinned the growth of 'just war' doctrines that were used to justify conflicts between Christian powers as well as with infidels of one kind or another. Although over the past two hundred years inter-state competition has had a political–economic cast, and in recovering the elements of the argument this chapter necessarily adopts an appropriately political–economic tone, the origins of justification for military and other competitiveness in the religious history of Western Europe suggest that it has a much more deep-seated cultural history that can only be touched on here.

The problem with the social argument for inter-state competition and the emergence of Great Powers is not only theoretical, as the next section attempts to show. It is also empirical. It no longer offers the same purchase on reality that it once seemed to. At present, a dynamic of globalization is prizing open even such previously closed or self-sufficient national economies as those of China, Russia and the United States, welfare states and government ownership are being undermined by privatization all over the world, and businesses increasingly operate in terms of world-regional and global markets (trends outlined in Chapter 4 in the discussion of states and power). In this historical context the imperatives of territorial and economic competition between states make even less sense than they did when economic and political practices jointly encouraged them (as they did from the late nineteenth century until 1945; see Chapter 6).

The purpose of the rest of this chapter is to identify the historical–geographical conditions in which the pursuit of primacy did perhaps make some sense and why today this component of the geopolitical imagination increasingly does not. Before we examine something of the historical geography of primacy, however, it is important to identify the major axioms upon which the presumption of a permanent or trans-historical pursuit of primacy rests.

Axioms for the pursuit of primacy

The hierarchy of Great Powers can be seen as emerging out of a competition for primacy on the basis of two axioms concerning states and their attributes. One is that relative power differences between states cause states to compete with one another for relative shifts in power and status. The second is that competition between states takes place under conditions of international anarchy; that is, conditions in which there is little or no return to cooperation, and winning is everything. These axioms have been adopted most formally and consistently by modern international relations theorists. But they also implicitly inform the modern geopolitical imagination as it has developed and changed down the years.

The focus on one's 'own' state and its security vis-à-vis the pre-emptive activities and potential depredations of others reflects the profound ontological insecurity (sense of loss of predictability and order) of people in the modern world. The geopolitical imagination has offered a reassuring response. Once security was no longer vested in a transcendental religious order with earthly enforcers, such as the medieval Christian Church, a substitute had to be found. Beyond the boundaries of the modern state, hypothesized such thinkers as the early modern Machiavelli and Hobbes, and the later Hegel, at least in typical readings of them (which is what matters here), lay an anarchic 'state of nature'. The geopolitical simplification of the world into 'friendly' and 'dangerous' spaces provided a practical means of giving order to this threatening and dangerous world. Sometimes a religious vocabulary has served to validate secular geopolitics in classical transcendental terms. Thus, one's most important opponent takes on a satanic cast (the United States as the 'Great Satan' for Iran's Ayatollah Khomeini, the Soviet Union as the 'evil empire' to US president Ronald Reagan) or stands for the Antichrist (as in American-Christian fundamentalist understandings of the Soviet Union during the Cold War). This rhetoric, however, is usually a mask for the sense of a mysterious and alien threat coming from distant shores (and, after the film *Independence Day*, from other planets?). Anarchy 'out there' can be countered only by making sure that 'out there' does not come 'here'. Only primacy, a dominant global position underwriting national security, can guarantee that it stays 'out there'.

The first axiom for the pursuit of primacy refers to the idea that the power of individual states grows and declines at different rates, largely as a result of differences in rates of economic growth. In this construction, as some states gain power, others are losing it. Power is always gained or lost relative to others, rather than possessed or not possessed in an absolute sense. This presumes a global pool of power of fixed volume that when drawn on by one state at an increasing rate diminishes that available to others. The international relations scholar Robert Gilpin (1981, 13) expresses this static storage-bin view of state power as follows: '[T]he differential growth of power of various states in the system causes a fundamental redistribution of power in the system.' The outcome, the historian Paul Kennedy (1987, xxii) asserts, is that, historically, 'relative economic shifts heralded the rise of new Great Powers which one day would have a decisive impact on the military/territorial order'.

Three behavioural assumptions must all be valid for relative shifts in economic growth to both make and shuffle the Great Powers. First, governments must see relative growth advantages as leading to improvements in the status of their states within the state system. They must orient themselves to achieving rates of growth that outdo their competitors. Second, rather like stereotypical capitalists accumulating capital, states are never satisfied with small increments of status or even being first among equals. They always aspire to be number one or 'top dog'. Third, rising power is seen as invariably leading to increased military obligations and international commitments. As commitments expand beyond the capacity of a Great Power to finance them, primacy dissolves as adversaries with more modest commitments take their place at the head of the international league table. In sum, uneven rates of economic growth are seen as having cumulative effects that trigger the 'rise and fall of the Great Powers', to quote Kennedy's (1987) phrase for the way in which primacy is first achieved and later undermined.

Much of the public discussion of Japan in the United States in the 1980s and early 1990s rested on these premises. The increased economic strength of Japan relative to that of the United States was seen as portending a shift in orientation from a trading state to a political–military one, thus meaning that Japan would steadily replace the United States as the world's number one Power. That Japan's economy was still only half the absolute size of the US one excited little or no interest. Japan had been growing faster than the United States (though by the early 1990s this was not the case) and this trajectory supposedly boded ill for the US position. US relative decline related to the growth and presumed military expansion of Japan, therefore, rather than change in absolute capacities or change relative to rates of economic growth in the American past.

'Sporting' metaphors have been important in both pointing to the 'essential' differences between national potentials for world-class performance and in suggesting analogies that appeal to popular common sense, a significant aspect of mobilization for war and other aspects of inter-state competition. Michael Shapiro (1989, 70) points out that comparing world politics to sporting contests serves the geopolitical purpose of emptying world space of any particular content: places lose their uniqueness and world politics becomes a type of strategic calculus, pure and simple. The world thus becomes like a gigantic playing field in which geographical location is equivalent to strategic location on a football field or basketball court. US presidents (particularly Presidents Nixon and George H. W. Bush) have been particularly fond of sports metaphors as applied to world politics. 'Punting', 'play selection' and 'end-run' were three American football favourites that President Nixon often applied to foreign policy. They allowed a notoriously socially awkward man to appear as 'one of the boys', engaging in dialogue with other sports-loving men. They also point to the obvious social-psychological linkages between sports as a type of sublimated war and the history of sports as a preparation for warfare. The Duke of Wellington may have started the trend in sports metaphors when he reportedly said that 'The Battle of Waterloo was won on the playing fields of Eton.' But, more critically, sports analogies allow a depoliticization of world politics. They encourage a sense of inter-state competition as a form of *pure* contest without moral or legal safeguards.

Zbigniew Brzezinski's book *Game Plan* is used by Shapiro as a convenient example of the genre. Supplying an 'integrated geostrategic framework for the conduct of the historical American–Soviet contest' (Brzezinski 1986, 8), the former national security adviser to President Carter implies in his use of a sports-contest discourse that

> what is simply involved is a struggle between two implacable forces that manifest a difference in ideology and historical motivation. He consistently represents the United States and the Soviet Union as 'two powers . . . fundamentally different'. And much of the book is addressed to aspects of this difference between the two contestants.
>
> (Shapiro 1989, 92)

The two contestants are depicted thus: one the 'real' man, the other the ersatz and lesser man. This obscures, of course, the costs imposed on others by the contest and the historical origins and interests served by continuing the contest into the indefinite future (Brzezinski's assumption at the time of writing). The image of contestants or competitors, therefore, naturalizes a process of global conflict between inevitable adversaries:

> Within the flattened narrative of the security contest, only national-level goals are implicated, and the 'threats' and 'catastrophes' that emerge as possible events touch only the interests of abstract entities situated in a geopolitical arena. The shape of the modern sports discourse and its use as a vehicle to figure world politics has the effect of situating us as spectators in a contest rather than as subjects in structures that create identities and locate us generally in a political economy distributing forms of danger.
>
> (ibid., 93)

The second axiom, that of international anarchy, derives from the view, implicit in the modern geopolitical imagination, that the state system is inherently conflictual at all times and in all places. In other words, if any one state gave up its ambition to be top dog, another would be sure to replace it. This is why relative power is given such weight. If one state increases its relative power, then another state must either try to imitate its success or balance against it by allying with other states. The former is a so-called hegemonic strategy, the latter is a balance-of-power strategy. In either case a 'unipolar' situation is seen as always momentary. Sooner or later, states seeking upward mobility will act to reduce their subordination. The fact that Napoleon Bonaparte, Kaiser Wilhelm, Hitler, General Tojo and Saddam Hussein seemingly acted on this basis to change the global status quo gives the perspective a set of persuasive historical examples for which prudence demands a response in kind.

The general perspective on international anarchy presumes, however, a universal condition of anarchy unaffected by space, time or reflection. Anarchy is the outcome of spontaneous self-serving action by states. From this point of view, anarchy cannot be seen as the result of the projection of a set of particular values that underwrite particular relations of power and authority. It is more like a market among states, in which each state is akin to an individual person pursuing its self-interest (whose definition is exogenous to the model) but devoid of any self-reflexivity about why it is doing so. It differs from the typical market, however, in that spontaneity in this case does not give rise to a collective benefit and order but to inequality, hierarchy and disorder.

This outcome results not only from the assumption of spontaneous action as states collide with one another but also from the lack of differentiation among both types of polity (ways of organizing political life) and types of state system (ways of organizing states) that have existed historically. From this perspective, all states are presumed to be alike and part of the system of territorial states originating in Europe in the seventeenth and eighteenth centuries (the Westphalian system, named after the Treaty of Westphalia of 1648, which was based on the idea of single rulers for clearly bounded territories mutually recognizing one another's sovereignty). But there has been a whole variety of ways of organizing power geographically over historical time (many coexisting with one another during the years of the Westphalian system itself), from hierarchical but territorially mobile units such as nomadic empires (e.g. the Mongols) to non-hierarchical but territorially fixed ones such as sedentary tribes (e.g. Scottish or Pashtun clans) with the most modern (European) units combining hierarchical and territorial features, further differentiated one from the other by the relative importance of capital accumulation and coercion by dominant classes in their political organization (city-states, nation-states and empires). There have also been distinctive types of state system in which different sorts of relationships have predominated between member units. The anarchic Westphalian system is the only type of system with presumed like-units jockeying with one another for advantage. Other systems have included:

1   suzerainties in which one unit has a superior status over others, e.g. US hegemony in its sphere of influence (the 'Free World') after the Second World War;
2   complex systems of hierarchical subordination and overlapping authority, e.g. the medieval European system in which the pope shared power with a variety of monarchs and feudal rulers; the United States before the Civil War (what Deudney (1996) calls the 'Philadelphian system' in which the states had much more power than they did after the Civil War); and the European Union; and

3 classical systems such as that of ancient Greece, involving heterogeneous but autonomous units.

Such a variety of polities and state systems mandates against presuming a single type of polity or state system as historically inevitable.

The assumptions of spontaneity and of absence of functional differentiation among types of state and state system are equally crucial to the presumption of international anarchy. The 'inevitability' of hierarchy is veiled by the mask of spontaneously generated anarchy among like units. This covers up ethical commitments that, as a consequence of the presumption of 'spontaneity', need never be acknowledged. International anarchy's basis in spontaneity serves to define a world of states that is the product of forces over which people have little or no conscious control. Inter-state competition just happens. Ironically, the purpose of the modern geopolitical imagination has always been to identify a rational basis for intervention in world politics even as world politics is construed deterministically. If intervention is possible, then there is also considerable scope for changing the values upon which intervention is premised. The assumption of inevitable anarchy is one of these.

The policy prescriptions to draw from the axiom of international anarchy have never been very clear, suggesting that its putative 'realism' is far from assured in all historical circumstances. Using the example of the contemporary United States after the Cold War, one reading of the situation would be that US governments should act to encourage multipolarity and a balance of power. US geographical isolation, the large national economy and possession of a nuclear first-strike capability guarantee the US a critical balancing role in a multipolar system. A different reading, however, would be that the United States should act to reassert its primacy before it is too late to do so. From this viewpoint, multipolarity is seen as inherently unstable as other states attempt to achieve hegemony by replacing the United States at the top of the international hierarchy. In both cases the axiom of anarchy offers no guidance.

There is an interesting analogy between the claim for a struggle for primacy among states and the 'winner-take-all' character of the reward structure of US society. In winner-take-all markets tiny differences in performance translate into massive differences in rewards. The relentless stress on 'winners' and 'blockbusters' in athletics, show business, publishing, management and universities (!) concentrates wealth and can encourage talented people into jobs that are less socially productive than other occupations they might pursue. Job markets are structured this way partly because of the multiplier effects of celebrity (record, advertising and book deals, for example, follow highly rated relative athletic or acting performances), but also because of the obsession with being number one or coming first – an important measure of status in a society in which there are relatively few widely accepted indicators for social stratification such as prestigious ways of speaking, widely accepted 'old boy' networks, etc. Rather like performers in a capitalist marketplace, states have been seen in the same light: engaged in a struggle for positional primacy that produces high returns for the successful few, but costly arms races and unproductive investments for all.

Although the obsession with coming first is such an obvious feature of US culture, it is by no means unique to it. The competitive element in the European state system has followed the extension of that system into the rest of the world, even if the mix of polities has been greater than the modern geopolitical imagination presumes. The rewards of success have appealed to political élites around the world, given that they are the ones who are most likely to benefit from enhanced international status. The winner-take-all analogy, however, has directly fuelled the recent national competitiveness debate in the United States. This implies that the future course of competition will be economic more than military. Again, states, like individual persons or business firms in a market, are seen as engaging in a struggle for survival in an anarchic world in which utility depends not on pay-off per se but on performance relative to that of others. In other words, the world of states is one of negatively interdependent preferences. Lower rates of aggregate economic growth relative to other Great Powers (such as Japan or China) are read as indicating an imminent loss of primacy, just as, in the past, falling behind in the arms race was given a similar reading. This focus on relative gains and losses characterizes much of the recent controversy in the United States over the 'threat' from Japan. In a story of almost legendary proportions, Robert Reich (Secretary of Labor in the first Clinton administration) had asked a number of groups of students, professional economists, investment bankers, senior State Department officials and citizens of the Boston area which of the following scenarios for the US economy was preferable: (1) one in which the US economy grows by 25 per cent over the next ten years, while that of Japan grows by 75 per cent; or (2) one in which the US economy grows by 10 per cent, while the Japanese economy grows by 10.3 per cent. Nearly everyone except the economists chose (2). Perhaps the magnitude of the difference drove some people away from (1). What is clear and disturbing, however, is that most people were willing to forgo a much larger absolute increase in economic growth to prevent a larger relative advantage accruing to Japan.

The idea of world politics as a horse race or sporting contest, in which finishing second is seen as finishing nowhere, has a persisting appeal. Indeed, US government economic policies are inspired by the same conception of relative gains. For example, US policies towards Japan on the export and manufacture under licence of satellites and the FSX fighter aircraft in the 1980s were developed explicitly in terms of relative gains and losses. But, again, this is not simply some quirk of US culture. The obsession with relative gains from inter-state competition is integral to the modern geopolitical imagination. It has a worldwide constituency of political élites. So, both as a representation of 'how the world works' and as a practice among politicians, the pursuit of primacy has a continuing importance in the conduct of world politics, notwithstanding the shift from military to economic competition as the presumed underlying dynamic.

The historical geography of primacy

Earlier chapters claim that the modern geopolitical imagination emerged in the historical context not only of Europe's overseas expansion but also of the progressive institutionalization of the European state system. In medieval Europe there were few fixed boundaries between political authorities. Communities were united only by feudal allegiance and personal obligation rather than by abstract conceptions of individual citizenship in a geographically circumscribed territory. As the identification of citizenship with residence in a specific territorial space became the central fact of political identity, sovereignty shifted from the person of the monarch to the territory of the state and its institutions. Although the Peace of Westphalia of 1648 is typically invoked as marking the beginning of the modern state system in which rulers recognized one another's right to rule over a given state territory, sovereignty is perhaps better thought of as a 'practical' fact within different territories that was only slowly recognized as a legal doctrine applying across all territories. Certainly, the idea of sovereign equality was not widely accepted until much later, and even then with all kinds of caveats and exceptions (think of the role of the Security Council in the United Nations Organization, for example). Notwithstanding its limited systemic effect, new claims to territorial sovereignty progressively undermined older forms of rule in Europe such as those of the Catholic Church and the Holy Roman Empire. Concurrently, the emerging capitalist economy was given regulative boundaries by the pre-existing, if still weakly centralized, territorial states. Not surprisingly, territorial states came to be widely seen as the exclusive sources of 'activity' on the world map. Industrial capitalism reinforced this tendency because its dominant spatial division of labour was strongly organized on a state-territorial basis. In particular, the stock Hobbesian view of individual persons locked in a struggle for survival in a hostile environment was easily extended to states. In this construction, the territorial state could be seen, as it is in the political economy of mercantilism, as a single abstract individual located in an environment of global anarchy.

The connection between this increasingly state-centric world and the modern geopolitical imagination lies in the sharp division that is drawn between the 'civic order' prevailing within state boundaries and the struggle for power between states beyond them. Just as major political ideologies provide the competing frameworks for political competition within states, the geopolitical imagination has provided the frame of reference for organizing state activities beyond them. Places near and far can be classified and related to the hierarchy of states: where are the main competitors? Where are the best possibilities for enhancing competitive position? Where are the main challenges to the established hierarchy? Maps of threats and dangers can be drawn based upon the inter-state pecking order at any particular time, identifying both its weakest and its strongest links.

Not all states have had the same capacity to realize their geopolitical visions. The so-called Great Powers of every era have been able to inscribe their particular geopolitical imaginations on to the world as a whole. State power has involved the capacity for undertaking action by states in the conditions of a particular era. This presupposes conventions of meaning about state behaviour that are shared by all parties and arise out of the actions of states and other actors. In this construction, *hegemony* refers to the norms and rules governing world politics accepted by dominant social groups and classes. The meanings are those diffused by the most powerful states; sometimes they emanate from single hegemonic states, indicating the material reality of primacy in certain historical settings, such as that of Britain in the period 1815–75 and that of the United States after the Second World War (see Chapter 6). But there have been periods, such as that from 1875 to 1945, in which the dominant representations and practices of states have not had a single origin or sponsor but have been more widely based among political élites in a range of states. Changing economic and technological conditions are of vital importance in determining the ebb and flow of candidacies for primacy (either in the sense of territorial or economic dominance over other states or as the primary 'rule-writer' in world politics). Particularly significant has been the relative ability of different states over time to capture the positive externalities from new technologies with novel economic and military applications.

One scheme (see Chapter 6 for more details) identifies three periods from 1815 until 1990 in which distinctive combinations of economic, technological and political factors produced very different conditions for primacy. The first (1815–75) rested on (1) a balance of power among a set of national states and territorial economies in Europe (plus the growing United States) with (2) an expanding tendency towards investment and trade organized non-territorially outside Europe with British institutions as the main organizing agents. The second period (1875–1945) was one in which a number of imperial rivalries produced a struggle for primacy by means of territorial control over chunks of the world economy. It is in this period that the transcendental 'urge' of Great Powers to subjugate one another could be said to have been practised in its most open form. Initially, the competition between imperial states (such as Britain and France) led to the creation of specialized colonial economies. These were organized largely around exclusive zones tied to particular Great Powers. This gave rise to enmity from those states (such as Japan and Italy) closed out of the enterprise. Failures in economic regulation at state and global levels (manifested in the Depression of the 1930s) encouraged a further retreat into territorial blocs. The final outcome was a struggle for supremacy between competing ideologies over how best to organize and, on one side, move beyond the blocs into an alternative type of political–economic organization.

The third period (1945–90) saw two states emerge as the main contenders for global primacy: the United States and the Soviet Union. The US agenda was for the United States to perform as an 'international' state preventing the re-emergence of the blocs that many influential Americans (and others) believed produced or exacerbated both the Depression of the 1930s and the Second World War. Three strategies were pursued to further this agenda. The first was the demilitarization and economic reorientation of Japan and Germany. The second was the containment of the Soviet Union and its state-socialist model of economic development through military alliances (such as NATO) and political–military intervention (as in Korea and Vietnam). The third was the creation of a set of institutions to project American practices and ideas about political–economic organization at a global level. These included the United Nations, the General Agreement on Tariffs and Trade, the International Monetary Fund, the World Bank, and such international agreements as the Bretton Woods Agreement of 1944 governing international monetary relations. By the 1960s these strategies had produced the beginnings of a globalized world economy in which many states became progressively internationalized. The costs of global military confrontation and the failure of the Soviet economy to keep up in information and communication technologies worked with other factors (such as social movements) to undermine the Cold War order. Pax Americana, designed to promote economic interaction between non-socialist national economies, encouraged a degree of globalization that effectively dented the powers of almost all territorial states, including the United States itself and its main adversary, the Soviet Union.

The pursuit of primacy, therefore, is not simply an ideological construct that has been 'exposed' as such by the recent destabilization of political boundaries at the end of the Cold War period. It does seem to have provided a socio-economic purpose to statehood under certain conditions. But it is difficult to accept it as a timeless attribute of world politics. In the first place, the pursuit of primacy has taken very different forms in different historical periods. The mix of military and economic elements, for example, has varied significantly over time, partly as a result of the shifting possibilities for mercantilist and open-trading economies. More importantly, the way the most recent period has ended with the erosion of many state powers may signal the beginning of its demise. What now seems increasingly clear is that as a feature of the modern geopolitical imagination, the transcendental conception of the pursuit of primacy involves the projection of a historically specific set of practices and representations on to world politics in general. With the value of hindsight, the two geopolitical assumptions that underwrote the plausibility of the pursuit of primacy as a fundamental element of the modern geopolitical imagination appear now as historically specific rather than generally valid: (1) that power flows from advantages of geographical location, size of population and natural resources combined in a territorial mode of production; and (2) that power is entirely an attribute of territorial states that attempt to monopolize it in competition with other states.

In the light of recent trends there are five specific ways in which the pursuit of primacy fails to pass a 'test' for trans-historical significance. First, a focus on relative power makes sense only in a world in which there is a constant threat of a war of attrition between Great Powers. In such circumstances constant preparation is required to achieve or maintain a relative advantage. The possession of and ability to deliver second-strike nuclear weapons render this vigilance obsolete. What matters in the nuclear era is not relative advantage – it matters little (how many times over can you destroy them?) – but absolute ability to inflict deadly damage (see Chapter 4). A single nuclear device planted in a major city by a 'weak' state (or stateless terrorists) could provide such an advantage without the immense cost associated with developing new weapons and their costly delivery vehicles. Indeed, in this situation a focus on relative advantage is dangerously misleading: it could give rise to a self-fulfilling prophecy as preparing for nuclear war makes such war more rather than less likely. The advent of nuclear weapons, therefore, irrespective of the political character of governing regimes (democratic/authoritarian, etc.), reverses the logic of war-preparedness first made in the modern era by Machiavelli. More need no longer be better. Nuclear weapons appear to make governments cautious, whether or not they are the United States or the Soviet Union (stateless terrorists are another thing entirely). The United States is in the situation (in 2000) of accounting for 50 per cent of world military spending, seven or eight times that of Russia, France, Germany and Britain, the other military Great Powers. By one calculus this is absolute primacy. But even those proposing continued high levels of military spending, such as the US Secretary of Defense, cannot find military threats that justify them. Only North Korea, Iran, Iraq and Syria are singled out as potential adversaries. Other Great Powers, presumably those most likely to be potential primates, do not figure at all in threat assessments. Other Great Powers are finding other things to do with their money.

Other features of contemporary military organization and technology are also making relative power less significant. One is the increased role of partisan or terrorist fighters in world conflicts. These are often stateless warriors, battling for this or that cause but without the infrastructure or the typical goals of a state-based military machine. Relatively simple but nevertheless devastating technologies (bombs, fuel-laden civilian aircraft, Stinger missiles, etc.) provide small mobilized groups with the means to neutralize or confound much more powerful states (consider the IRA in the United Kingdom, the ETA in Spain, and *al-Qaeda* around the world, for example). Another is the access to the information networks that are at the heart of military preparedness. Not only can small states and non-state groups potentially gain entry to and destroy command-and-control systems (if they have talented hackers and a good computer), they can also engage in 'netwar' (as have the Zapatista rebels in southern Mexico): using the Internet to send messages to sympathizers and non-governmental organizations so as to publicize and undermine state counter-insurgency operations. This is information warfare with a new twist, one that benefits small marginalized groups more than the big battalions of the Great Powers.

Second, economic growth and prosperity among the Great Powers have been enhanced since 1945 by increasing international trade and investment. The focus on relative gains and losses reflects a mercantilist economic ideology in which national economies are seen as equivalent to containers that can be filled only at one another's expense or through territorial expansion. Today, rates of economic growth are much higher when economies are open rather than closed. This is one of the lessons of the great Soviet collapse of 1990–92. Access to global markets is now the principal precondition for sustained economic growth. In turn, increased economic interdependence increases the incentive for resolving disputes non-militarily. The assets of businesses and individuals from particular states are no longer tied exclusively to the state's territory. Such foreign ties and stakes are a force for symmetrical interdependence when they are shared by all parties. This reduces the element of conflictuality and increases the return to cooperation.

One obvious criticism of the competitive primacy model, of course, is that states are not at all like competing firms in an economic model of perfect competition. For one thing, as noted previously, they are not all alike, whereas competing firms often are. Neither is it clear that most states actually have single goals analogous to increased market share or greater profits. They (like persons and firms) can also benefit from cooperation with one another. Military alliances and regional trading arrangements are examples of ways in which such benefits have long been realized. Another criticism, made by advocates of the advantages to all parties involved of free trade and comparative advantage, is that national economies often suffer when restrictions are placed on imports and exports. In this view, although special interest groups concerned about the prospects of this or that industry may like to persuade everyone otherwise, entire national economies are *never* in substantive competition with one another *when there is relatively open trade* (for an argument along these lines, see the economist Paul Krugman's (1995) reply to the historian Paul Kennedy (1995)). There are returns to specialization that can be captured only through either firm organization (e.g. vertical integration) or local/regional concentration. As an important trend in the world economy is away from the former to the latter in leading economic sectors (e.g. informational technologies, aerospace, producer services), national economies (as opposed to more localized ones) are even less likely to be in 'competition' than they were in the past when industries were more organized on a national rather than a transnational basis. An ironic confirmation of this trend can be found in comparing the title of the management guru Michael Porter's best-known book, *The Competitive Advantage of Nations* (1990), with its contents. The book is in fact about the increasingly localized and regionalized basis to economic performance within industrialized states; about what geographer Allen Scott (1996) calls the new 'regional motors' of the world economy.

Symptomatic of the internationalization of states is the growing concern of both states and firms with 'market access' (as described in Chapter 4). Though states continue to control the rules and structures of their own national economies, through formal international agreements and informal but explicit bargains they also grant access to foreign competitors. Classic free trade and investment policies, therefore, are being extended into the areas of industrial policy and domestic regulation to facilitate cross-border business collaboration. This is an important consequence of the recent rounds of the World Trade Organization (WTO) and the growth of regional trading blocs such as the European Union and the North American Free Trade Agreement (NAFTA).

Two measures of this growing internationalization are the increasing difficulty of accounting for trade in terms of transactions between national economies and the growing number of products without clear national origins. The tremendous expansion of foreign direct investment and subcontracting means that the national trade balances for countries such as the United States and Japan contain within them large numbers of items manufactured overseas by firms operating from within their confines. For example, according to the official books, the US trade deficit in 1986 was $144 billion. But if you include the activities of US-owned firms abroad and foreign-owned firms in the United States, the huge deficit disappears and becomes a surplus of $77 billion. The two figures for Japan are closer, reflecting the lower rate of foreign investment by Japanese industrial companies (Table 5.1).

What is now traded is also less often final products than components, materials and specialized research, design, fabrication, advertising, and financial services. One example comes from that quintessential US company General Motors, whose well-being was once considered to parallel exactly that of the United States as a whole:

> When an American buys a Pontiac Le Mans from General Motors, for example, he engages unwittingly in an international transaction. Of the $10,000 paid to GM, about $3,000 goes to South Korea for routine labor and assembly operations, $1,850 to Japan for advanced components (engines, transaxles, and electronics), $700 to the former West Germany for styling and design engineering, $400 to Taiwan, Singapore, and Japan for small components, $250 to Britain for advertising and design services, and about $50 to Ireland and Barbados for data processing. The rest – less than $3,000 – goes to strategists in Detroit, lawyers and bankers in New York, lobbyists in Washington, insurance and health care workers all over the country, and to General Motors shareholders all over the world.
>
> (Reich 1991, A6)

Table 5.1 *Conventional and ownership-based trade balances, United States (1986) and Japan (1983), in US$ billions*

		United States	Japan
Exports	[1]	224	146
Less intra-firm transfers		123	60
Plus local sales to foreign MNCs		267	3
Plus sales by home-owned MNCs abroad		777	150
Equals total foreign sales	[2]	1,145	239
Imports	[3]	368	114
Less intra-firm transfers		191	65
Plus local purchases from foreign MNCs		445	58
Plus purchases by home-owned MNCs abroad		446	90
Equals total foreign purchases	[4]	1,068	197
Conventional trade balance	[1] – [3] =	–144	+32
Ownership-based trade balance	[2] – [4] =	+77	+42

Source: Adapted from Julius (1990, 81)

Note: MNC = multinational corporation

A third failure of the pursuit-of-primacy thesis is that US hegemony has been qualitatively different from that of previous eras. It has been institutionalized globally through a large number of agencies and has had a profound cultural influence. The demise of the Soviet Union has left the United States as the only Great Power with a global message: mass consumption, personal liberty, private property, markets and electoral democracy, to name just the most obvious elements of this message. Of particular significance, US efforts at enmeshing other states into international 'regimes' of one kind or another covering a wide range of substantive issues from trade to fishing rights and environmental degradation have had the effect of emphasizing collective more than unilateral approaches to resolving conflicts.

Fourth, the number of truly Great Powers is now very small and the incentives for cooperation between them are very high. Under conditions of interdependence, each state has a stake in seeing that the others adhere to common rules governing trade, security and investment. A diffusion of power or pluralism, more than a multipolarity among singularly 'selfish' states, is now the emerging trend among the Great Powers. This points up the social nature of international relations. Conventionally, and reflecting the *idée fixe* of the pursuit of primacy, relations between states are seen as inherently conflictual, as in the Prisoner's Dilemma game (Figure 5.1). In this game, both actors have identical preferences, both acquisitive and precautionary. Each must act without information concerning how the other will act. Each can benefit without making the other worse off by cooperating (+20 apiece). But the greatest return comes from defecting before the other does (+40). If both opt for defection, both experience losses (−20 apiece). Without learning and confidence-building measures this is the presumed likely outcome. In this game a two-party conflict will not produce cooperation since each party believes it can always get more by not cooperating (see the higher potential returns to non-cooperation than to cooperation; potential losses are, of course, commensurately higher). This presumes, however, that each state is a self-sufficient actor without divided interests, that there is a set of restrictive informational constraints and that there is a lack of institutional and communication channels for accommodating differences of interest. Recent experience of international negotiations calls each and every one of these presumptions into question.

Fifth, and finally, there is a growing élite and popular consciousness of global interdependence. Slowly and fitfully, a sense of a common global fate has started to grow. This is encouraged by the spread of access to information about previously unknown and distant places. But it is more directly the result of a growing sense of linkages and dependencies in a more interdependent world. This need not translate into an immediate bonding with neighbours, let alone distant others. It does indicate, however, the arrival of a consciousness of the impact of distant events (oil spills, burning forests, refugees from communal strife, nuclear winter, etc.) on the prospects for life at 'home' that do not have simple military solutions to them. The success of transnational activists in Europe and North America in generating sanctions against apartheid South Africa in the 1980s is a good example of how both a non-state-based movement and a concern for norms of conduct by states rather than national interests are now an important part of world politics. The recent upsurge in ethnic conflicts might suggest exactly the opposite. Indeed, there obviously are situations where group identities and interests are so polarized and threatened that peaceful resolution of conflict is next to impossible. But new spatial practices involving increased trade and population movement can be expected slowly to force a sense of common fate even as they also generate enmity and

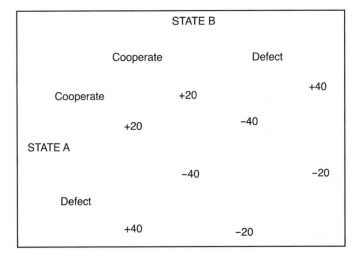

Figure 5.1 The Prisoner's Dilemma game. In this game both actors have identical preferences, both acquisitive and precautionary. Each must act without information of how the other will act. Each can benefit without making the other worse off by cooperating (+20 apiece). But the greatest return comes from defecting before the other does (+40). If both opt for defection, both experience losses (−20 apiece). Without learning and confidence-building measures this is the presumed likely outcome.

Source: Author

conflict. Some parties to the Israel–Palestine dispute, undoubtedly one of the world's most intractable conflicts, increasingly see a future in which each side will have to depend on resources and skills provided by the other. Why? Because what neither side can do 'is conjure up a pair of coherent nation-states' (Avishai 1995). Each population is now too intermingled with and interdependent on the other.

Beyond the pursuit of primacy

In the contemporary world, therefore, the pursuit of primacy has limited possibilities. Of course, this world might not last. Elements of opposition to the disruptive consequences of globalization are increasingly visible. The emerging transnational liberal order can be seen as detracting from rather than augmenting the material conditions of life for both the 'average' and the poorest people in countries where this process has gone the furthest. The 'hollowing out' of the US economy, due more to technological change and the globalization of production than to the direct effects of trade, has produced a deepening inequality of incomes between both social strata and regions. This could lead to increased success on behalf of calls for a reterritorialization of the US economy through protectionist measures (such as tariffs, quotas, content requirements and product liability regulations). The calls are already there, whether from those concerned about immigration or those worried about the lack of jobs for the unskilled and the uneducated.

Unfortunately, in the world economy as it is now organized, this would probably only lead to greater disinvestment from the territorial economy of the United States. This is what happened in France in 1983 when the Mitterrand government nationalized assets and put restrictions on banks and was immediately faced by a massive outflow of capital. What is more likely to meet with success is if societies collaborate multilaterally to regulate the socially disruptive consequences of globalization. Just as in the past, social democratic movements arose to counter at the state level the socially damaging impacts of unfettered markets, so in the future, 'multilateralism will become an arena of conflict between the endeavour to buttress the freedom of movement of powerful homogenizing economic forces, and efforts to build a new structure of regulation protecting diversity and the less powerful' (Cox 1992, 177). Whatever happens, therefore, a return to the pursuit of primacy seems an unlikely recipe for anyone's success, except those interests in the militaries and the defence industries whose careers and profits have come to depend on it. The world in which states must now operate has changed beyond recognition within one generation. A variety of economic and political processes have encouraged this. Prime among them have been the globalization of production and finance, the informational revolution in computing and telecommunications, the internationalization of states enmeshed in regimes of all kinds, massive migrations from poorer to richer countries, and the spread of a 'world culture' emanating from the Hollywood film studios, Tokyo video game manufacturers, Milan fashion designers, and MTV.

The most vital geopolitical consequence of the displacing of state boundaries by the flows of people, goods and capital moving between local nodes in global networks has been the undermining of conventional processes and understandings of spatial hierarchy. The pursuit of primacy presupposes a world map in which state territories can contain within their boundaries the lion's share of the most profitable transactions that they generate. The returns from these transactions can then be invested in political–military adventures of one kind or another designed to establish the state's primacy or high ranking among 'comparable' states. For much of the past two hundred years this has largely been the case. In this respect, the modern geopolitical imagination's commitment to the pursuit of primacy reflected the dominant practices of the Great Powers whose world it sought to understand and expand. US hegemony after the Second World War, however, undermined the pursuit of primacy in building the foundations of a world economy that has now escaped from the territorial confines of existing states.

The pursuit of primacy requires a hierarchy of settled political identities like those implicit in Hegel's model of lordship and bondage and in conventional understandings of the origin of the hierarchy of states. Changed historical–geographical conditions draw attention to this requirement. The trend from a geography of boundaries to one of flows that is integral to globalization, therefore, signals a crisis of representation for the modern geopolitical imagination. There is a crisis in the intellectual authority of established conventions about the behaviour of the Great Powers – the main active figures in classical geopolitics. Not the least of these is that of the pursuit of primacy. But how did the modern geopolitical imagination work in practice between Hegel's day and ours? That is the subject of Chapter 6.

Recommended reading

Agnew, J. A. 2001 Disputing the nature of the international in political geography. *Geographische Zeitschrift*, 89: 1–16.

Collins, R. 1986 *Weberian Sociological Theory*. Cambridge: Cambridge University Press.

Cox, R. W. 1992 Multilateralism and world order. *Review of International Studies*, 18: 161–80.

Croxton, D. 1999 The Peace of Westphalia of 1648 and the origins of sovereignty. *International History Review*, 21: 569–91.

Deudney, D. 2000 Regrounding realism: anarchy, security, and changing material contexts. *Security Studies*, 10: 1–45.

Ferguson, Y. H. and Mansbach, R. W. 1996 *Polities: Authority, Identities, and Change*. Columbia, SC: University of South Carolina Press.

Frank, R. H. and Cook, P. J. 1995 *The Winner-Take-All Society*. New York: Free Press.

Inayatullah, N. 1997 Theories of spontaneous disorder. *Review of International Political Economy*, 4: 319–48.

Julius, D. 1990 *Global Companies and Public Policy: The Growing Challenge of Foreign Direct Investment*. New York: Council on Foreign Relations.

Kahler, M. 1992 Multilateralism with small and large numbers. *International Organization*, 46: 681–708.

Kennedy, P. 1987 *The Rise and Fall of the Great Powers: Economic Change and Military Conflict from 1500 to 2000*. New York: Random House.

Krugman, P. 1996 *Pop Internationalism*. Cambridge, MA: MIT Press.

McNeill, W. H. 1982 *The Pursuit of Power*. Chicago: University of Chicago Press.

Mahan, A. T. 1965 *The Influence of Sea Power upon History, 1660–1783*. London: Methuen (original edition Boston, 1890).

Mearsheimer, J. J. 2001 *The Tragedy of Great Power Politics*. New York: Norton.

Porter, M. 1990 *The Competitive Advantage of Nations*. London: Macmillan.

Reich, R. 1991 *The Work of Nations: Preparing Ourselves for 21st Century Capitalism*. New York: Knopf.

Shapiro, M. J. 1989 Representing world politics: the sport/war intertext, in J. Der Derian and M. J. Shapiro (eds) *International/Intertextual Relations*. Lexington, MA: Lexington Books.

6 The three ages of geopolitics

The purpose of this chapter is to show how three specific epochs or eras of geopolitics developed over the period from the early nineteenth century to the 1980s, built upon the interaction of changing material conditions with the 'principles' of the modern geopolitical imagination described in previous chapters. Even though the geopolitical imagination of each 'age' is distinctive in many respects, there are continuities as old themes are recycled in new contexts. The three discourses or modes of representation I discuss are referred to as civilizational geopolitics, naturalized geopolitics and ideological geopolitics, respectively. World politics in each of the epochs has been organized around the characterizations of space, places and peoples defined by these modes of representation and communication.

Continuity and periodization in the modern geopolitical imagination

There is an obvious continuity running through the modern geopolitical imagination in a number of ways. One is the continuing use of a language of spatial difference expressed in terms of a temporal metaphor (modern/backward). Another is the world visualized as a whole as a field of reference for inter-state relations. A third is the exclusive role of territorial states as the actors in world politics. Finally, there is the pursuit of primacy by Great Powers as the motivating force behind world politics. However, the idioms and contexts of usage have changed dramatically over time. There is considerable intellectual danger in assimilating all geopolitical discourse into an overarching continuity flowing from the Renaissance to the late twentieth century and arbitrarily selecting themes or linguistic forms without attention to the contexts (historical and geographical) in which they have arisen. After all, the technological, economic and social conditions that constrain and enable both representations and practices have changed significantly over the years. For example, the world economy today is organized completely differently from how it was during the heyday of European colonialism in the late nineteenth century.

Obviously, any periodization is inherently contestable. Not only are the beginning and ending dates subject to criticism for false precision, history is itself dynamic and not readily divided into neat periods. Periods, therefore, are easy to hate. This is particularly so if they are established, as they inevitably are, from the specific geographical and social location of a particular observer, or seen as 'stages' with one leading inevitably to the next in a progression towards historical destiny. But trying to understand the course of history means imposing some sort of order on it. This is where periods come in. Periods threaten understanding only when we forget that they are imposed on a more complex flow of history. Otherwise they are extremely useful; indeed, they are one of the few ways analytically to order an otherwise endless flow of events. Literary historian David Perkins (1992, 64) well captures both the importance and the discomfort of periodization when he writes of periods that they are

> necessary fictions ... because one cannot write history or literary history without periodizing. Moreover, we require a concept of a unified period in order to deny it, and thus make apparent the particularity, local difference, heterogeneity, fluctuation, discontinuity, and strife that are now our preferred categories for understanding any moment of the past.

A periodization of geopolitical discourse, therefore, obviously simplifies a more complex flow of representations and practices. Each period has within it the seeds both of its own demise and of subsequent periods. The dating of periods is tentative and contestable. Each could be subdivided to represent the inevitable shifts and turns in understandings and practices. Different places do not necessarily share all the features that each period is taken to represent. For example, the political élites of different Great Powers might have different emphases from one another. Subordinate or colonized peoples would not necessarily share the cultural and economic presuppositions upon which the modern geopolitical imagination relied at any particular time in any specific place. The issue periodization addresses, however, is that of dominant trends, understandings and associated action, and how these differ over time. Such a historicist conception does not deny the continuities identified in previous chapters or the differences from place to place. Rather, the idea is that in each period the various 'principles' or foundations intersect in different ways to produce a different mix of geopolitical representations and practices. This gives rise to different hegemonies, dominant sets of rules governing world politics, in different epochs (see Chapter 1). These correlate with the economic, technological and social trends noted in Chapter 5 as giving rise to the threefold periodization: first, the early nineteenth century (1815–75); second, 1875–1945; and third, 1945–90. But it should go without saying, given the arguments and claims of earlier chapters, that the dominant geopolitical imagination flowing through these periods has older roots in the growth of state formation and capitalism in early modern Europe and European expansion into the rest of the world. Neither are the hegemonies intrinsic to the different eras free of contradiction or contest. Indeed, it is out of their internal contradictions that old hegemonies are undermined and new ones arise.

Civilizational geopolitics

In the late eighteenth century a civilizational geopolitics emerged as part of the reaction to the struggle for stability in a Western Europe that had lost its cosmic centre based on a common Christianity at the time of the Wars of Religion. This was an element in what Toulmin (1990, 170) has termed the attempt at constructing 'a more rational Cosmopolis, to replace the one lost around 1600'. This version of the modern geopolitical imagination peaked in influence in the first half of the nineteenth century. Before that time, its various principles were not engaged with one another in a common whole in the way they were to be after 1815. The international political economy from 1815 until 1875 was characterized by a European Concert in which no one state 'laid down the law' for the others within Europe, and by an emerging British economic dominance in much of the rest of the world. The Concert was designed to limit the revolutionary impulses emanating from the French Revolution of 1789 and its aftermath. Wars to foment revolution or to acquire territory were seen as threats to the balance of power between the dominant states or Great Powers. At the same time, Britain's head start in economic growth as a result of its Industrial Revolution produced a different interaction with the other continents as compared with that of the other European states (save for the Netherlands). The search for markets for manufactured goods and the availability of capital for overseas investment produced pressures for external expansion (both colonial and more generally) that set Britain apart. As the rest of Europe became caught up in national economy making, a process largely completed in Britain, British businesses and political leaders created a worldwide network of trade and financial flows.

At this stage in the growth of world capitalism, British centrality was widely seen by European political élites as serving not only a national interest but a global interest. The British national economy became the locomotive of the world economy. The recycling of capital from worldwide investments made London the world's main financial centre. The main political interest of financial capital lay in preventing economic disruption in places where it had investments. Not surprisingly, British governments also came to value stability and often intervened militarily outside Europe when the status quo was under threat. With the increasing economic success of the German and US economies, however, the British economy, still largely committed to older technologies, was faced with a dilemma. On the one hand, it was faced with increased competition in manufactured goods and, on the other, it had a relative advantage over the others in its access to the rest of the world. Consequently, after 1875 Britain turned away from Europe and the United States (where British capital had major investments) towards its empire and those world regions where its hegemony was apparently more secure. This gave impetus to the final collapse of the Concert of Europe, already threatened by the emergence of new states in Europe (such as Germany and Italy) without any commitment to it, and the emergence of a global system of competing empires as others, from Germany, France, the United States and Japan, attempted to follow in Britain's footsteps. But it was within Europe that the most important challenge of the age arose. The political order of civilizational geopolitics was that defined by aristocratic and bourgeois élites. In its main features it was designed to tame and repress the nationalism generated by the French Revolution. But as the nineteenth century wore on, a new conception of statehood based on the creation of nation-states came to the fore. These were states built upon cultural divisions and particularities. In the identity of state with nation, territorial sovereignty was fused with the fate of the nation. The interests of peoples were rigidly territorialized as the number of states in Europe tripled. This new system of states had no place for civilizational geopolitics.

It was in the political–economic context of 1815–75 that civilizational geopolitics was most fully established, even though the principles it drew on had older roots, as described in previous chapters. Its main elements were a commitment to European uniqueness as a civilization; a belief that the roots of European distinctiveness were found in its past; a sense that though other cultures might have noble pasts with high achievements, they had been eclipsed by Europe; and an increasing identification with a particular nation-state as representing the most perfected version of the European difference.

The idea that the Earth's land area is divided into separate continents was first proposed by ancient Greek geographers, who identified three continents – Europe, Asia and Africa – bounded by the Mediterranean Sea and the Rivers Nile and Don (see Chapter 2). Although later geographical knowledge suggested that Europe and Asia were not clearly demarcated from one another by a significant body of water, the division of the world into continents persisted because the concept of Europe itself changed. From a physical-geographical region the geographical realm of Europe was transformed into a cultural region. This happened as the Christian Church abandoned its claims to universality and defined a much more narrowly circumscribed Christendom. The Arab and, later, Ottoman 'perils' gave particular credibility to the sense on the part of vulnerable Europeans of a profound chasm between the familiar world of Christian Europe and the exotic world of the Muslim Other (see Chapters 2 and 3).

Imaginative maps showed Europe as a 'queen' among the continents, complete with orb and sceptre (Plate 6.1), a portrayal that illustrates not only a sense of difference but also an emerging sense of superiority. On the one hand, the monarch represents dominance but, on the other, the female body represents the fertility and flux associated with ocean-based imperial expansion. This imagery was reinforced by the European voyages of discovery, which demonstrated the self-evident initiative, vision and zeal of Europeans. Over the next centuries, the feeling of superiority 'gradually hardened into an inflexible conceit that held Europe to be the most civilized and best governed of all the world regions' (Bassin 1991, 3).

The basis for this conceit was not simply comparison with benighted non-Europeans but, more importantly, a sense of a distinctive past (described in Chapters 2 and 3). The rest of the world was 'available' for use by Europeans because their history destined them for Greatness. It is no mere coincidence that the early nineteenth century saw an obsession among European élites with the examples of ancient Greece and Rome. This was particularly marked in the United States – in the minds of its leaders very much an extension of European ideals beyond the physical boundaries of Europe. The founding fathers of the United States looked to Rome for their model; they thought of Athens as ruled by mobs. In the early nineteenth century this view was displaced by a much more popular Greek revival that affected everything from political oratory to town names and house design. Throughout Europe a classical education became a prerequisite for social and political success. Contemporary events were understood in terms of classical referents.

The dichotomy drawn between Europe and the other continents was reinforced by the frequent combination of an 'original' homeland with a colonial periphery or frontier which characterized such different states as Spain (in the Americas), England (in North America), the United States (beyond the eastern seaboard) and Russia (in Siberia). In each case, however, they had a colonizing experience closer to home. The Spanish had the south they had retaken from the Moors; the US coastal settlement was itself the result of invasion. The British Empire, for example, could be said to have begun in Ireland and elsewhere in the Celtic fringes of the British Isles (as the film *Braveheart*'s portrayal of English invasions of Scotland in the period 1290–1314 reminded its twentieth-century viewers). Colonialism started in Europe. It was later extended across the Atlantic and then to other world regions. In Ireland the rulers of England had their first major experience of subjugating, mapping and renaming a land occupied by an (intermittently) hostile but (always) 'inferior' people. That this process was incomplete because of persisting religious and political differences continued to concern the British élite down into the nineteenth century. The wild and defiant image of the Irish remained at odds

Plate 6.1 Europe as a queen, in Sebastian Munster's *Cosmographia Universalis* (1588). This conveys both the symbolic sense of Europe's difference and its aristocratic superiority.

Source: Basle University Library

with the civilizing mission of the English: to bring order and civilization to its oldest and closest colony. The first Others, therefore, were not abstract and unknown but concrete and familiar.

Beyond the limits of the homeland, the *noblesse oblige* (the obligation those of high rank have to those lower down the social order) that aristocratic élites felt towards their social inferiors was projected more generally. At best, the political ideas of most European imperialists were that, as in the case of the British in India,

> political power tended constantly to deposit itself in the hands of a natural aris-
> tocracy, that power so deposited was morally valid, and that it was not to be

tamely surrendered before the claims of abstract democratic ideals, but was to be asserted and exercised with justice and mercy.

(Stokes 1959, 69)

This idea of a natural aristocracy led the British rulers to ally themselves with their local counterparts, who, notwithstanding their racial difference, were seen as their natural collaborators. They were rarely disappointed. In this context, learning about and understanding the workings of colonized peoples, particularly their social hierarchies, was a vital part of the burden of spreading the light of European civilization into the world's darkest corners. The Nigerian novelist Chinua Achebe (1975, 5) captures the challenge and how it was thought of as follows:

> To the colonialist mind it was always of utmost importance to be able to say: I know my natives, a claim which implied two things at once: (a) that the native was really quite simple and (b) that understanding him and controlling him went hand in hand – understanding being a precondition for control and control constituting adequate proof of understanding.

Though obviously paternalistic and denigratory, this conception of others did incorporate a sense of wonder and even awe about the world into which Europeans had ventured. Indeed, many Europeans saw a nobility and simplicity in the lives and beliefs of the natives that they encountered. Analogies with ancient Greece and Rome often underwrote this attitude, even down to the images used to represent religious rituals and performances (Plate 6.2).

Unlike in the case of the European empires proper, in which the European 'motherland' was separated from most of its colonies by large tracts of water, in the United States and Russia there was no such clear separation. Peripheries and frontier zones could be identified but there were not always obvious physical boundaries. The problem was resolved for a time in the Russian case by designating the Ural Mountains as the specific boundary between the European and Asian parts of the empire. Indeed,

> [t]he basic geographical proposition that Russia divided cleanly and naturally into Asian and European sectors entered into the very foundation of the imperial ideology that was refined in the course of the eighteenth century. It was disseminated in the geography texts that began to appear in ever-greater numbers after 1750 and by the end of the century it had become a universally accepted truism.

(Bassin 1991, 7–8)

In the United States, as European settlement spread into the continental interior during the early nineteenth century, the expanding frontier became the root and symbol of the filling out of national territory to the Pacific. Divine Providence was invoked initially to justify the invasion of the continent. The dominant justification, however, became the providential mission of 'America' to spread American ideals and institutions to the Pacific – and beyond. The rhetoric of mission peaked in the 1830s and 1840s when zealots wrote of American 'Manifest Destiny' and 'the Great Nation of Futurity'. In

Plate 6.2 'Indians from Canada make a sacrifice to the Great Spirit' from Bernard Picart, *Cérémonies et coutumes religieuses des peuples idolatres* (Amsterdam, 1735). This drawing shows the wonder and simplicity (and classical analogies) that eighteenth-century Europeans saw in primitive religion.

Source: Research Library, The Getty Research Institute, Los Angeles

1847 the US Secretary of Treasury placed in his annual report a section referring to the aid of a 'higher than any earthly power' that had guided American expansion in the past and that 'still guards and directs our destiny, impels us onward, and has selected our great and happy country as a model and ultimate centre of attraction for all nations of the world' (quoted in Pratt 1935, 343).

As this last quotation suggests, the agents of civilizational geopolitics, while invoking a general European outlook and agenda, came from different states that competed with one another as claimants to the mantle of the ancients. 'New Romes' abounded. Yet the claims rested more upon the imitation of French and English models of nation building than upon some universal shared experience of the 'classical' European past. From successfully policing religious affiliations, these prototype modern nation-states had expanded their jurisdictions to construct a set of collective interests, outlooks and traditions.

Particular dynastic and regional identities were reworked as *national* cultural identities. Classical motifs were combined with local ones. The Napoleonic armies that swept across Europe in the aftermath of the French Revolution left architectural and organizational residues of its classical hubris, from the triumphal arches to the Code Napoléon. But a second influence emanating from England and the German states soon spoke for a less remote past:

> Gothic churches, vaulted tombs, museums and houses of assembly, adorned with mementoes of medieval battles and national heroes, filled the lacunae of the nation's collective memory, and little children were taught to revere Arthur and Vercingetorix, Siegfried and Lemminkainen, Alexander Nevsky and Stefan Dusan as much [as], if not more than, Socrates, Cato and Brutus.
>
> (Smith 1991, 81)

The particular national genius was thereby highlighted against the achievements of a common but ancient past.

This harking back to medieval roots served to foster the sense of membership in singular communities with distinctive attributes that demanded 'self-realization' in 'historic homelands'. The French and English pioneers of nationhood, however imperfect or partial their own achievements, provided important role models for other Europeans, from the Greeks battling the Turks to the Germans and Italians struggling with their historic geographical divisions (described in Chapter 3). Not surprisingly, it was to them, and particularly the Britain that emerged victorious from the Napoleonic Wars, that the new nation-states also turned for recognition of their status as modern nation-states.

A clear distinction was drawn between the possibility of this new statehood in Europe and its impossibility elsewhere. Within Europe (and wherever Europeans now lived), nations and states could be conjoined as limited territorial entities that balanced one another mechanically in a system of weights and balances. This analogy to Newton's mechanics provided the image of stability within which foreign policy between nominally equal states was possible. Outside this system was an unlimited space of primitive or decadent political forms that were candidates for conquest rather than recognition. The 'boundary' example of the Ottoman Empire provides a case in point. It was not recognized as a 'member' of the Concert of Europe until 1856, and even then the long-run disputes between it and Russia were never given the attention they deserved. The

Otherness of the Turks was a fundamental barrier to their participation in a civilizational geopolitics that drew hard lines around its European homeland and even had trouble including such 'marginal' Europeans as the Russians and Americans.

Naturalized geopolitics

In cultural anthropology the idea of *totemism* refers to the practice of interpreting nature in the vocabulary of human groups, such as kinship and genealogy, whereas *naturalization* does the opposite. It represents the human entirely in terms of natural processes and phenomena. This is what happened to geopolitics from the late nineteenth century down until the end of the Second World War, even though it had older roots (as indicated in previous chapters) and persisted in some respects thereafter (see the next section on ideological geopolitics). Rather than being a feature of civilization, geopolitics was now largely determined by the natural character of states that could be understood 'scientifically', akin to the new understanding of biological processes that also marked the period. The understanding of states and the geographical organization of the world economy that this gave rise to underpinned the inter-imperial rivalry of the geopolitical order from 1875 to 1945.

As the old Concert of Europe unravelled in the 1870s, two antagonistic groups of states emerged. One, led by Britain and France (with covert US support), stood for the coexistence of open trade and imperialism. The other, headed by Germany, was revisionist – concerned to build their own empires and challenge British financial dominance. By the 1890s this division had become the major feature of world politics. Its basis lay partly in the dramatic growth of the German economy but also partly in the intense nationalism of the period. National economy making through the extension of national railway systems and the reorganization of economic space to accord greater centrality to national markets and finance was combined with the spread of literacy in vernacular languages and universal elementary education to produce an enhanced sense of national difference and exclusivity on the part of mass publics. Nationalism was no longer just for élites. In imitation of one another and as an outlet for investment during the economic depression of 1883–96, a massive expansion of European empires brought much of the world that had remained outside the reach of European imperialism into a geographically specialized world economy. Different world regions took their place within a global division of labour, as plantation agriculture brought bananas to Central America, tea to Ceylon (now Sri Lanka) and rubber to Malaya (Plate 6.3). The main axis of capital accumulation lay in the home–colonies structure of the British Empire and the US–British relationship. The challenge for Germany and the other revisionists, such as Italy and Japan, was to build an alternative to this structure. Ultimately that proved impossible without taking on the British and their allies militarily. The climax arrived in 1914 when the alliance systems on either side of the inter-imperial rivalries set in motion in the 1870s came to life in response to the seemingly minor conflict over the political status of Bosnia-Herzegovina.

The apparently inexorable and inevitable slide into war associated with the First World War was the outcome of a mindset on all sides that saw war as the only way out of the political–economic impasse into which they had all slipped. The conduct of the war itself was also the outcome of a mentality in which the individual person had to sacrifice for the good of the greater whole: the nation-state that defined the totality of one's identity. Humanity had lost control of its destiny. Nature ruled in affairs of state. The naturalized geopolitics characteristic of this epoch had a number of telltale features: a world divided into imperial and colonized peoples, states with 'biological needs' for territory/resources and outlets for enterprise, a 'closed' world in which one state's political–economic success was at another one's expense (relative ascent and decline), and a world of fixed geographical attributes and environmental conditions that had predictable effects on a state's global status.

Naturalization had a number of preconditions. One was the (apparent) separation of the scientific claim from the subject position of the particular writer or politician. Claims were made to universal knowledge that transcended any particular national, class, gender or ethnic standpoint. So, even as a particular national interest was self-evidently addressed, the act of doing so was framed by a perspective that put it into the realm of nature rather than that of politics. A vital aspect of this, and another precondition for naturalization, was the conviction that observation of the world's political and economic divisions was a form of innocent perception from which generalizations about resources and power could be deduced. The pre-existing commitment to this view of perception and its associated mapping, naming, and citing of spaces as 'colonial', 'European', 'powerful' or 'backward' was not often acknowledged.

'Scientifically', in the empiricist sense of the term, the world beyond the immediately familiar was blank and empty but was then filled and labelled according to its varying natural attributes as they appeared from Europe. The world was then known and possessed not just politically but epistemologically. This was the great achievement of

Plate 6.3 'Jungles to-day are gold mines to-morrow': advertising poster of the (British) Empire Marketing Board, September–October 1927

Source: Constantine (1986)

naturalization: to have depoliticized inter-imperial rivalry into a set of natural and deter-mining geographical 'facts of life'. The invention of political geography during this period is of a piece with this trend. On the one hand, the new field was openly touted by its advocates as a way of educating both élites and national populations in the ways in which physical geography constrained and directed state formation and empire building. On the other hand, it based its claims to seriousness on offering a perspective that was useful primarily for the nation-states from which its advocates came. Thus, Halford Mackinder provided a global geopolitical model (see Chapter 2) but was concerned primarily about the implications of this model for the future of the British Empire.

There were, of course, conservative, socialist and anarchist currents of thought that tended to decry the emerging dominion of naturalized modes of thought and action. But even they increasingly tended to flirt quite openly with claims to scientific status that would put them beyond politics and into nature, so to speak. The voluntarist under-standings of nationhood and class that had emanated from the American and French Revolutions were increasingly marginalized. For example, the polemic of Ernest Renan (1882) against the German 'blood and soil' conception of nation on behalf of a volun-tarist one (of choice or will) was rapidly eclipsed.

One of the most important elements in naturalized geopolitics was the distinction between imperial and colonized peoples. Citizenship laws, classifications of colonies (particularly between European settler colonies and others), religious missionary activities and even the emerging division of labour between university subjects (e.g. sociology versus anthropology) operated on this distinction. But this distinction was not merely a recognition of necessity – a pragmatic response to an obvious spatial division. Ironically, it rested on the view that (some) Europeans had become masters of nature as a result of superior 'fitness' in a natural process of evolution. Transforming or displacing more primitive peoples was justified because, as Charles Darwin (1839, 520), one of the intellectual sources for this type of reasoning, had put it, 'the varieties of man seem to act on each other in the same way as different species of animals – the stronger always extirpating the weak'.

In the United States in the 1880s, 'scientific' racial claims largely replaced provi-dential/civilizational ones as the basis for American destiny. Anglo-Saxons (and other 'Teutons') were identified as exemplars of superior evolutionary fitness. Their superior political principles, growth in numbers, and economic power were evidence enough (notwithstanding contradictory evidence in the census of the United States, which suggested the greater growth of other groups). One leading proponent of the fitness argument saw an Anglo-Saxon future. 'The day is at hand', he wrote, 'when four-fifths of the human race will trace its pedigree to English forefathers, as four-fifths of the white people of the United States trace their pedigree today [1885]' (quoted in Pratt 1935, 348).

The principle of natural selection thus filtered down from scientific theorizing into popular culture largely in terms of the idea of the 'survival of the fittest'. This became a staple of the journalistic accounts and travel writing that accompanied the great explo-sion of European imperialism at the turn of the century. An older idiom justifying colo-nialism in terms of moral uplift and religious grace gave way to a discourse of racial competition and dominion. Initially invoked to distinguish the major 'races' of humanity, it quickly lent itself to more refined distinctions, as with the American 'Anglo-Saxons', and served as an important inspiration for the racist ideologies – racial anti-Semitism, anti-Slavism – and eugenics programmes that flourished politically in Europe and North America in the 1920s and 1930s. These ideas were shared internationally rather than associated singularly with any particular country such as Germany. For example, to recall the example of England and Ireland from the previous section on civilizational geopolitics, the late nineteenth century saw a racialization of the presumed inferiority of the Irish. To the historian Lord Acton the whole language of English success and Irish failure as imperialists was attributable to the natural inferiority of the Irish race:

> The Celts are not among the progressive . . . races but those which supply the materials rather than the impulse of history, and are either stationary or retrogressive. . . . They are a negative element in the world . . . and waited for a foreign influence [the English] to set in action the rich treasure which in their own hands could be of no avail.
>
> (quoted in Curtis 1983, 53)

Within Europe, Jews were especially vulnerable to the growth in 'scientific' racism. Religious anti-Semitism had deep roots in Europe going back to the Middle Ages. But not only did the nineteenth century bring rapid advancement to Jews as a group as they moved out of ghettos into the wider society, but some non-Jewish groups 'experienced rootlessness, fragmentation, and estrangement from a once securely anchored, familiar world' (Barnouw 1990, 79). As a result, Jews became objects of hatred and fear from those nostalgic for a world they had lost but without good prospects in the new national–industrial society. Demagogues readily portrayed Jews as rootless cosmopolitans or allied with the nation's enemies in a European world dividing into parochial nation-states (Plate 6.4). There was no longer a space for ethnic and racial difference within the boundaries of states. Jews were dangerous polluters of national homogeneity and disrupted a presumed natural order in which racial groups and states are mutually dependent. Every 'race' was seen as requiring its proper place. Jews were increasingly portrayed in 'racial' terms. In the work of the nineteenth-century political geographer Friedrich Ratzel (1844–1904), the founder of the ecological theory of race, the Jews are seen as the one race most 'out of place'. Sander Gilman gives the flavour of Ratzel's linkage of place with race when he writes:

> In the Near East they [Jews] were productive (for example, creating monotheism) but in Europe they have no real cultural meaning. The association of place and race is linked in the rationale of the German in Africa or the Jew in Europe. They are presented as mirror images, for while the German in Africa 'heals', the Jew in Europe 'infects'.
>
> (Gilman 1992, 183–84)

The Nazi geopoliticians of the 1930s came up with formalized schemes for combining imperial and colonized peoples within what they called 'pan-regions'. However fanciful in terms of possibilities of overcoming the political–economic relationships of the time, such mappings did express in extreme form the common assumption that the world was constituted by racial groupings that could be neatly divided into two 'types' of peoples. The one essentially existed to serve the other. Dominant and subordinate races were joined together territorially in the pan-regions. Whatever the precise influence of the Nazi geopoliticians, particularly Karl Haushofer, upon the practices of the Nazi regime, there is little doubt that their ideas fitted into a larger political context in which notions of racial hierarchy were conjoined with conceptions of state 'vitality' to justify territorial expansion. The *Zeitgeist* (spirit of the age) even drew into its orbit intellectuals with impeccable anti-naturalist credentials. The classic case in point is the German philosopher Heidegger (1959, 39), who wrote in 1935, apparently without tongue in cheek: 'We are caught in a pincers. Situated in the center, our Volk incurs the severest pressure. It is the Volk with the most neighbors and hence the most endangered. With all this, it is the most metaphysical of nations.'

At the same time, either through analogy or literally, the European territorial state acquired a status as an organism with its own 'needs' and 'demands'. This too was a transposition from evolutionary biology. But it had older roots. German idealist philosophers of the early nineteenth century such as Hegel (1770–1831) and Fichte (1762–1814) had regarded the state as a being or entity with a life of its own. It was but a short step from this to the idea of the state as an organism, a step facilitated by the spread

Plate 6.4 'Behind the enemy powers – the Jew': Nazi racial propaganda

Source: © Bildarchiv Preussischer Kulturbesitz, Berlin

of biological ideas into the emerging social sciences and the rhetoric of politicians in the 1890s.

Like all 'organisms', a state must struggle against the environment (in this case, other legitimate states and 'empty spaces') to survive. This need to struggle required that it acquire space and resources to feed its healthy growth. The rebirth of militarism in the late nineteenth century after its decline in the aftermath of Napoleon was a corollary trend that further fed the image of the state as a permanently embattled entity that could prosper only if individuals and classes subordinated their particular interests to the interest of the larger whole. The doctrine gained expression in terms of three central points: the harmony of state and nation, natural political boundaries, and economic nationalism.

The late nineteenth century was a time of tremendous social disruption in Europe. Not only were there massive movements of people within Europe and across the oceans leading to an explosion of urbanization, but increasing capital mobility also undermined the national and local circuits of savings and investment that had previously given an appearance of long-term commitment to the activities of capitalists. These trends were particularly strong in Britain, where by the 1890s a reaction had set in against their social consequences. One was the growth of political movements pushing for political and economic rights for workers and women. Another, increasingly popular with political élites, was directed at nipping these new movements in the bud by 'turning the clock back' to a time when all social groups were thought to have lived in local social harmony – only this time, the social harmony was to be realized in the conjunction of the nation with the territorial state.

A mythic medieval community in which each social stratum knew its place and duty to the totality was projected on to the nation as a whole. It was this re-creation of the local past in the national present that served to give the state its organic character. An important contrast was drawn between this and the workings of a laissez-faire economy. Such an economy was seen as undermining the organic unity of the nation-state. To some influential conservatives it was seen (and still is seen) as a source of decay, what Oswald Spengler (1926) was to term after the First World War the source of 'the decline of the West'. The founders of disciplinary geopolitics, the Swede Kjellén and the Englishman Mackinder, both subscribed to this viewpoint.

This perspective was widely shared by emerging political élites throughout Europe (and its overseas extensions). In Russia the Urals were increasingly discounted as a geographical divide. One influential intellectual current came to see Russia as 'a transcendental geo-historical, geo-political, geo-cultural, geo-ethnographical, and even geo-economic entity, designated by a new generic term: *mestorazvitie*' (Bassin 1991, 16). In Germany local patriotism or identity with the *Heimat* (or homeland) was channelled by conservative political parties into an overarching identity with the Reich. Only through the advance of the Reich could the *Heimat* be defended. In the United States, until the Civil War 'the United States' was typically a plural noun. Afterwards, it became singular. This transformation is symbolically associated with President Lincoln's famous Gettysburg Address of 1863 when 'What had been a mere theory of lawyers like James Wilson, Joseph Story, and Daniel Webster – that the nation preceded the states, in time and importance – now became a lived reality of the American tradition' (Wills 1992, 145).

Another element in the view of the state as an organic entity was the idea that a state had 'natural boundaries'. This implied, first of all, that the historical boundaries were not necessarily the proper ones. The territorial status quo of the Concert was now called into question. But it also implied that all the members of a putative nation or ethnic group had a natural right to live within the boundaries of the state. Finally, it also opened up the possibility of using natural features to designate the natural area of the state. Swedish conservatives, such as Kjellén, argued against Norway's independence in 1905 partly because they claimed that the Scandinavian mountains were not a natural boundary. The Nazi concept of *Lebensraum* (borrowed from the German geographer Ratzel), justifying German territorial expansion in Mitteleuropa (and elsewhere) and what was actually called the 'intellectual liquidation' of Poland (Burleigh 1988, 50), had their roots in the notion of natural boundaries.

This logic was never extended to the 'stateless' colonized world. This is most clearly illustrated in the case of Africa. In the wake of the exploration of Africa, conquest and colonial rule came quickly and devastatingly. In 1884–85 the major European powers agreed at the Congress of Berlin to stake out their spheres of influence in Africa. The 'Scramble for Africa' that followed over the next twenty years produced lines on a map which had little relation to underlying cultural or economic patterns. Elsewhere, the establishment of colonial boundaries was less hasty if often no less arbitrary. These designations continue to haunt these regions to this day (see Chapter 4).

Lastly, a major tenet of 'organic conservatism' was economic nationalism. The state was seen as defining the basic unit for economic transactions. Firms and individuals were held to be subordinate to the greater needs of the nation-state. This too was biological in nature. Writers as different in other respects as the English economist Hobson (who influenced Lenin's thinking) and the English geographer–politician Mackinder both shared organic definitions of national interest as the driving force behind economic growth. To Hobson, empire sacrificed the 'home' economy, while for Mackinder, empire was the means of maintaining the economic basis of the military power that was essential for national survival. What they and others would have agreed on was the organic unity of the national economy as a 'going concern'. The German *Weltpolitik* (roughly, economic imperialism), though often at loggerheads with doctrines of *Lebensraum* or territorial expansion, nevertheless provided a similar popular logic for acquiring overseas markets and sources of raw materials to underpin German industrial success.

The idea of a 'closed world' was vital to the plausibility of the language of biology for understanding the European state as an organic entity. As frontiers closed in North America and the world's land masses were incorporated into the world economy, control over territory appeared to be a crucial prerequisite for economic growth. For British élites in particular, the certainties of a world economy that worked in their favour seemed to be gone. They were faced by both protectionist rivals and a domestic commitment to free trade that the new geopolitical order had made anachronistic. 'The British economic lead had evaporated, and for the British the world did indeed seem to be shrinking, to be closing in' (Kearns 1993, 29).

But there was more to it than this. From 1880 to 1914 'a series of sweeping changes in technology and culture created distinctive new modes of thinking about and experiencing time and space' (Kern 1983, 23). Such innovations as the telegraph, the telephone, the car, the cinema, the radio and the assembly line compressed distance, truncated time and threatened social hierarchies. The global spread of railways and the invention of the aeroplane were perhaps the most important challenges to conventional thinking about time and space. At much the same time the mapping of the oceans proceed apace. The principles of scientific navigation established by such figures as the American

Matthew Fontaine Maury (1806–73) transformed the oceans from trackless hazards into a network of highways with well-charted currents and wind paths (Hearn 2002). The sense of a closed world, therefore, was neither illusory nor the product of a uniquely British sensibility.

In a 'closed world' a premium would be placed on relative national efficiency. States must therefore organize themselves to increase their productivity relative to their rivals. No state could afford to rest on its laurels:

> Nationalism and protectionism helped countries mobilize the resources of their earth and their people. . . . A country without access to the full complement of modern industries was vulnerable, would be a pushover in a war, and thus would attract the bellicose attention of more well-balanced nations.
>
> (Kearns 1993, 18–19)

The major political dispute in late nineteenth-century/early twentieth-century Britain concerned the merits of maintaining the policy of unilateral free trade introduced in 1870. Both sides adopted the language of national interest and 'shared a belief in the importance of "national economic power" but they lacked agreement on exactly what the concept meant or how it should be measured' (Friedberg 1988, 79).

If the liberal economics of the nineteenth century came under attack from the 1890s on, it suffered a nearly fatal final blow from the Great Depression of the 1930s. The mass unemployment of the time produced a variety of intellectual and political reactions. State corporatism pushed the organic analogy to its limits. From this point of view, ascendant in Fascist Italy, Nazi Germany, Spain and Portugal, old economic assumptions no longer applied. The national unit was now pre-eminent. State corporatism was 'a system of interest and/or attitude representation, a particular modal or ideal-typical institutional arrangement for linking the associationally organized interests of civil society with the decisional structures of the state' (Schmitter 1971, 86). The state could now be freed from guaranteeing such values as individual liberty and equality to pursue its 'own' agenda of security and foreign affairs with the total support of its economy.

The breakdown of the capitalist world economy also offered an opportunity for some form of socialist internationalism. But Stalin's adoption of 'socialism in one country' in a Soviet 'socialist motherland' effectively reproduced the territorial economism characteristic of other regimes rather than called it into question. This model of socialism – central economic planning, collective agriculture and new political élite – did threaten established élites, however, particularly where communist political parties affiliated with the Soviet Union were strong, and led to the association between socialism and 'subversion' that was to be so important in geopolitical discourse after the Second World War.

It was the rationale for government management of the economy provided by the English economist John Maynard Keynes that provided the most important justification for stimulating the national economy in Britain and the United States in the 1930s and 1940s. Liberalism remained important in both these countries when it had long disappeared elsewhere. Keynes provided a way of 'squaring the circle', by arguing for a synthesis of private and state activity through 'countercyclical demand management'. The national economy, however, was the basic unit for Keynesian macroeconomics just as it was for the other political–economic philosophies. Indeed, its basic measures such as national 'propensity to save', national investment and national productivity were indicators of national efficiency that earlier and more illiberal figures would have readily recognized for what they were.

The final feature of naturalized geopolitics was its emphasis on the determining character of geographical location or environmental conditions. The relative success of different states in international competition was put down to absolute advantages of location and to superior environmental conditions. 'Maritime' states (states on the edges of land masses) were seen by many military strategists as possessing intrinsic advantages over 'interior' states because they had fewer contiguous or neighbouring states, and hence fewer potential adversaries. Maritime or sea-power states were seen as outflanking 'continental' or land-power states in control over the oceans, the main means of global movement. Only the coming of the railway had called this view into question – and this because of the relative size or weight of the Eurasian land mass (or 'Heartland') in relation to the difficulty of coalescing and policing the 'insular crescent' (Mackinder) or 'rimland' (Spykman) around its edge. The use of the Mercator projection to represent the Earth's land areas served to exaggerate the sense of a world dominated by Eurasia (especially Siberia) because of its systematic enlargement of polarward areas relative to equatorial ones (see Figure 2.3 on p. 29).

This spatial-geographical determinism associated with formal geopolitical models, however, was never as popular as a less specific (and more ambiguous) environmental determinism. From this point of view the Great Power potential of states was a function of their industrial prospects, which in turn could be traced to their natural resources (particularly energy resources) and ability to exploit them. Some went further and claimed that this ability was itself determined by climate. From 1900 until the 1930s such views were not exceptional. Indeed, they formed the mainstream of opinion, particularly among those educated in such academic fields as geography, geology and biology. Much of the academic geography of the period in Germany and the English-speaking world involved elaborating systems of environmental/geographical accounting; classifying states and regions in terms of inventories of resources, racial characteristics, economic and political organization, and climatic types. These were taught in schools and became the conventional wisdom about why some places had 'developed' while others had lagged behind; natural attributes determined national destiny. The manuals issued to US soldiers during the Second World War are classics of this genre; their title is revealing of their content: *Geographical Foundations of National Power* (issued by the US Army in 1944).

The Second World War, therefore, was not external to this geopolitical discourse. Naturalized geopolitics pervaded the representations of the war itself in the film and cartographic propaganda that both sides engaged in. On the Axis (Germany–Italy–Japan) side it provided the logic for the war (expand or perish) and (in the German case) for the anti-Semitic Holocaust that was one of its central features. On the Allied side the war was seen as a struggle for survival by peaceful maritime states whose success was based on their capacities for invention and trade.

The outcome of the war brought to an end the geopolitical order of inter-imperial rivalry and created the conditions for the construction of a new post-war geopolitical order: 'The emerging order was a qualitatively distinctive one, characterized by the break-up of the old colonial empires through the decolonization process, and the emergence of the United States as an economic, military, and political hegemon' (Biersteker 1993, 16).

Ideological geopolitics

By definition, the modern geopolitical imagination is ideological, if ideology is defined as an amalgam of ideas, symbols and strategies for promoting or changing a social and cultural order – or, as the anthropologist Paul Friedrich (1989, 301) puts it, 'political ideas in action'. After the Second World War, however, the geopolitical imagination was centred much more explicitly around competing conceptions of how best to organize the international political economy. Cold War geopolitics was 'linguacultural' (Friedrich's term) more than civilizational or naturalized. By this I mean that the values, myths and catchwords drawn from the experiences of the two victorious states, the United States and the Soviet Union, were to define and determine the terms of the geopolitical imagination of the period. One of these, the United States, was to prove the more effective in gaining widespread acceptance for its model of political–economic organization. But its success relied heavily on the active presence of the other as a point of comparison and threat.

The ideological character of the conflict defining the period needs emphasizing. The historian Anders Stephanson (1995, 123) has noted that this is largely why the Cold War remained *cold*. The spheres of influence established immediately after the Second World War proved rather long-lasting because, given each side's possession of nuclear weapons that each could deliver to the homeland of the other, the party disturbing the status quo would face potentially devastating consequences. In response, the definition of ideologically opposite geopolitical zones put a premium on maintaining doctrinal purity on both sides of the 'fence' that divided the two worlds. Consequently,

> Geopolitical and military realities served to keep the struggle preeminently in the realm of ideology. The rivalry generated an arms race of mind-boggling waste and destructiveness, and it fueled numerous deadly conflicts in the Third World; but never once did it escalate into open hostilities between NATO and the Warsaw Pact. . . . The two sides 'found it convenient', if I may steal a phrase from Trollope, 'to establish a mutual bond of inveterate hatred'.
>
> (ibid.)

Under the dictatorship of Stalin from 1943 to 1947 the Soviet Union built a formidable military economy that required as its premise the existence of a major external threat. The recent invasion by Nazi Germany meant that a sense of threat was not difficult to sell to the Soviet population. Associated with this sense of external danger was an identification with regimes and revolutionary movements that were opposed to the US/Western model of economic development. On the US side the government set out from 1944–47 onwards to sponsor an international order of economic liberalism in which military expenditures would provide a protective apparatus for increased international trade. This would in turn redound to the advantage of American businesses and stabilize the American economy. The key institutions and rules spread rapidly in the immediate post-war period. All the major industrialized countries except the Soviet Union accepted the rules of the international economic game as written in Washington, DC, either through external inducement (as with the aid from the 1947 Marshall Plan) and coercion (as with the British dollar loan of 1946) or through direct intervention and reconstruction under US auspices (as in West Germany and Japan). Many of the elements of this system have lasted into the 1990s, although some faded away earlier. They are:

1   indirect stimulation of economic growth by means of fiscal and monetary policies;
2   commitment to a growing global marketplace based on a global division of labour;
3   accepting the dollar as the principal world currency;
4   hostility to Soviet-style economic planning;
5   assuming the burden of policing political changes that could be construed as damaging to the stability of the world economy.

In the end, domestic conditions in the two main states undermined the global status of each. By the 1980s the Soviet economy had failed to deliver both improved military equipment and improved living standards for the Soviet population as the Communist Party and its leadership sank into complacency. For the United States, internationalization was too successful. By the early 1970s the growth of other economies produced pressures on the dollar and the drain of policing proved increasingly expensive.

Not simply international relations but also the economic and political arrangements of all of the world's states beyond the Soviet sphere of influence (China for a few years in the early 1950s, Eastern Europe until 1989) came under the spell of the US system. The vital glue for this system was provided by the political–military conflict with the Soviet Union. Even during times of coexistence or détente, the overarching Cold War conflict served both to tie such important states as Germany and Japan into the US system and to define two geopolitical spheres of influence in which each generally accepted the other's dominance. This imposed a stability on world politics, since the United States and the Soviet Union were the two principal holders of nuclear weapons and they could threaten to escalate conflicts if their 'fundamental strategic interests' were threatened, even as they promoted numerous 'limited' wars in the Third World of former colonies where each armed surrogates or intervened themselves to prevent the other from achieving a successful 'conversion'. For all their political and economic weaknesses, however, many Third World countries had to be wooed and cajoled. Unlike in the previous period, the world map was no longer a vacuum waiting to be filled by a small number of competing Great Powers. The United States and the Soviet Union remained attached to anti-colonial philosophies even as they engaged in imperial competition. They had to win friends and influence people to extend their spheres of influence; they could no longer simply impose external rule.

In this setting, ideological geopolitics developed the following major characteristics: a central systemic–ideological conflict over political–economic organization; 'three worlds' of development in which US and Soviet spheres of influence vied for expansion into a 'Third World' of former colonies and 'non-aligned' states; a homogenization of global space into 'friendly' and 'threatening' blocs in which universal models of capitalism–liberal democracy and communism reigned free of geographical contingency; and the naturalization of the ideological conflict by such key concepts as containment, domino effects and hegemonic stability. Older themes from civilizational and naturalized discourses were worked into the new discursive space. The two most important have been the backward versus modern polarity and the idea of 'national security'. These will be mentioned as need arises.

The Cold War began as a series of US policies designed to rebuild Western Europe after the Second World War but became a system of power relations and ideological representations in which each 'side' defined itself relative to the Other that it was not. This happened in an ad hoc manner but had deep linguacultural roots.

The United States and the Soviet Union had been allies during the war. In the aftermath of the war a basic question concerned how Western Europe was to be organized politically and economically. US moves to spread American views of how this might best be done met with resistance from the Soviet Union and its allied communist parties in Western Europe. From 1945 to 1950 this initial conflict was

> reinforced by a crisis here, a Soviet move there, and an analysis of the protagonists [in the United States] which insisted that Moscow was impelled to expand and that only the United States could prevent it from achieving world domination.
>
> (Cox 1990, 30)

The 'loss' of China to communist revolution was a particularly important element in deepening US–Soviet mutual hostility.

But US and Soviet mutual suspicion had origins that went back to the Bolshevik Revolution of 1917. There had been a 'red scare' in the United States then that raised the spectre of domestic subversion. US military forces also intervened in the Russian Civil War against the Bolsheviks; the US government knew what it was up against. In the 1920s a fear that alien influences threatened the workings of the United States' unique institutions led to a resurgence of isolationism. 'Foreigners' were dangerous. The United States refused to recognize the new Soviet government in the old Russian Empire.

Both the United States and the Soviet Union were peculiar states. They both had origins in revolutions with explicit ideological agendas. They both claimed popular mandates that transcended particular ethnic, class or regional interests. They both offered themselves as uplifting lessons in political–economic experimentation to a world where cynicism was rampant. In this context, the practices of foreign policy took on a special importance. In two states where what precisely was either 'American' or 'Soviet' was unclear, the threat of things un-American or anti-Soviet became central to national identity.

Ideologically, neither saw itself as a limited territorial entity, though in each there had been long-standing disputes over the geographical scope of their respective revolutions. 'Isolationism' versus 'internationalism' has been the typical language of foreign policy debate in the United States. But by the 1940s the dominant élites of each aspired to global ideological dominion. In the Soviet case this was obvious in the ideological lineage that informed official ideology (Marx–Engels–Lenin), if less obvious in the official practice of isolating the Soviet economy from external linkages. In the United States the urge to export the American ethos was long-standing. Certainly, from the 1890s onwards 'America' itself was increasingly seen by leading Americans as an idea that could be sold:

> American traders would bring better products to greater numbers of people; American investors would assist in the development of native potentialities; American reformers – missionaries and philanthropists – would eradicate barbarous cultures and generate international understanding; American mass culture, bringing entertainment and information to the masses, would homogenize tasks and break down class and geographical barriers. A world open to the benevolence of American influence seemed a world on the path of progress. The three pillars – unrestricted trade and investment, free enterprise, and free flow of cultural exchange – became the intellectual rationale for American expansion.
>
> (Rosenberg 1982, 37)

This outlook had more idealistic origins. In 1823 President James Monroe had issued the statement that the United States would oppose all foreign intervention in the American hemisphere. This doctrine became as sacred to many Americans as the Declaration of Independence and the Constitution. Over the years, however, it served to justify US military intervention anywhere in Latin America and the Caribbean, even though this was to give it a meaning quite different from the original sentiment expressed by Monroe. During the Cold War period the Monroe Doctrine licensed US governments to intervene throughout the western hemisphere and, by extension, into the rest of the world. An originally democratic impulse was thus recycled under new conditions and in so doing undermined what it had originally stood for. Ironically, given the global pretensions of each, domestic subversion by foreign agents is a dominant theme in each state's view of its own history, as is an assumed vulnerability to external threats. Again, the perpetual threat to revolutionary achievements is probably of some significance. In the United States appeals to public opinion have also been of importance in legitimizing any kind of foreign policy. Hence, exaggerating vulnerability and invoking the need for 'self-defence' became an important means of mobilizing public opinion on behalf of foreign ventures. The arrival of nuclear weapons (and long-distance delivery systems) led to an American sense of connection to events elsewhere in the world, through the possibility of escalation of local conflicts into a global nuclear one, that previously had been missing. The long history of US territorial security has, however, perhaps left people without much of a basis for judgement about external 'threats', even as the Soviet experience of invasion and mass murder as recently as the Second World War provided a more obvious basis for collective paranoia.

One important consequence of this shared sense of vulnerability was an idealization by each of the other. Each became a super-potent adversary in the eyes of the other. This is particularly obvious in the US case, where there was systematic exaggeration of Soviet economic and military capabilities. For example, Holzman (1989) has shown systematic exaggeration in official US estimates of Soviet military spending from the 1960s to the 1980s. Gervasi (1988), in his annotated version of the Pentagon publication *Soviet Military Power*, finds evidence of massive overstatement of Soviet military capabilities – and we now know, after the collapse of the Soviet Union, that he was absolutely correct. As late as 1988, US foreign policy was still based on the continuing existence of a singular Soviet threat to the US position in the international political economy. Even as the Soviet Union disintegrated in 1988–89, a group of leading US experts on military and foreign affairs produced a report that assumed an indefinite continuation of Cold War bipolarity. Though the economic problems of the United States – relating to huge trade and federal government deficits – were well known, it ignores them. Instead, it engages in statistical and cartographic legerdemain to demonstrate *increased* US vulnerability to the Soviet threat. It is no exaggeration to say that political élites in each country became so obsessed with each other that, as one of their number put it, he hoped they would not appear to future historians as 'two dinosaurs circling one another in the sands of nuclear confrontation' (Gorbachev 1987–88, 494).

In the context of peculiar revolutionary states with quite different recipes for political–economic organization, abstract terms such as 'communism' and 'capitalism' took on culturally loaded meanings. As the United States came to personify capitalism, so did the Soviet Union represent communism. Each state became the geographical manifestation of an abstract political economy. Each was foreign and dangerous to the other. The constitutive divide in world affairs after the Second World War thus had specific linguacultural roots that reduced communication to a repetition of buzzwords about the nature of the other party.

Joanne Sharp (2000) has shown that in the American case after the end of the US–Soviet wartime alliance a widely circulating magazine like the *Reader's Digest* served up a steady diet of articles by 'experts' that created a 'common-sense' knowledge about the Soviet Union and the threat it posed to Americanness. Its readers were not unprepared. The *Digest* had been worried about the Soviet Union since the early 1920s. Now readers were bombarded with accounts of the total difference and opposing character of the Soviet Union as compared with the 'actually existing' United States, not simply differences in aspiration or historical experience. Each was masqueraded as the actual opposite of the other. Such representations had powerful effects in mobilizing public opinion into the Cold War consensus that progressively engulfed American politics from 1947 until the Vietnam War of the late 1960s.

Certain domestic interests were served by this geopolitical reductionism. Once under way, this reinforced its pervasiveness. In the Soviet Union it disciplined potential dissidents into support of a monolithic state apparatus. In the United States it produced a consensus around a 'politics of growth', an enlarged military economy, and opposition to any politics (usually construed as 'socialist' or leftist because of associations between socialism and the Soviet Union) that would subvert the country from within. In other words, it eroded the possibility of an open, competitive democratic politics. The very identity of being American became associated with a narrow political spectrum at home and a virulently anti-socialist/anti-Soviet position (they were rarely distinguished) abroad. 'The simple story of a great struggle between a democratic "West" against a formidable and expansionist "East"' became 'the most influential and durable geopolitical script of [the Cold War] period' (Ó Tuathail and Agnew 1992, 190).

As the 'leader' of the West, the US president played a key role in giving meaning to the Cold War:

> In ethnographic terms, the U.S. President is the chief *bricoleur* of American political life, a combination of storyteller and tribal shaman. One of the great powers of the Presidency, invested by the sanctity, history and rituals associated with the institution – the fact that the media take their primary discursive clues from the White House – is the power to describe, represent, interpret and appropriate.
>
> (ibid., 195–96)

Through the recycling and repetition of certain images and themes taken from the American past, the president can give a sense of continuity to the practice of foreign policy.

On 12 March 1947 President Truman laid an important piece in the groundwork for the Cold War geopolitical imagination in a speech to the US Congress that while it 'drew the line' against communism in Greece, it also integrated this action into American history by referring to 'free peoples who are resisting attempted subjugation by armed minorities or by outside pressure'. An American audience knew well that this was a reference to their own experience. Connecting the Cold War with the American Revolution was thereafter a central theme of Cold War discourse. President Reagan, for example, was to regard the Nicaraguan Contras ('anti-Communists') of the 1980s as the 'moral equivalents of the [US] Founding Fathers'.

After the Second World War the US political élite read three lessons from the war and the Great Depression that preceded it: that appeasement of potential aggressors was dangerous; that US national security depended upon the global balance of military forces and not just the balance in the western hemisphere; and that the United States must everywhere oppose attempts at building regional economic blocs or pan-regions such as had emerged in the 1930s. These became the persisting themes of Cold War discourse as they were incorporated into US government practice.

They led inevitably to a forward posture in confronting the Soviet Union. Europe was the main setting for challenging the possibility of Soviet expansion beyond the confines accepted by Roosevelt, Stalin and Churchill at the Yalta Conference in early February 1945. Both the United States and the Soviet Union came to share a mutual interest in this scenario. Berlin and its division became the symbolic centre of the Cold War, where each side confronted and stared down the other (Plate 6.5). The geographical division of Europe guaranteed US and Soviet status as the dominant partners in the alliances they formed in the two halves of Europe. Very quickly, the common goal became the maintenance of balance. 'In this sense the Cold War was more of a carefully controlled game with commonly agreed rules than a contest where there could be clear winners and losers' (Cox 1990, 31).

Plate 6.5 Dismantling the Berlin Wall: symbol of the Cold War after it was erected by the East German government in 1961 to physically divide the eastern section of the city from the Western/ capitalist West Berlin; its fall on the night of 9–10 November 1989 represented the symbolic end of the Cold War. Here people celebrate the end of the division of Berlin on the night in question at the Brandenburg Gate, the historic centre of the city.

Source: Author

Although there were episodic uprisings and movements against this spatial division, particularly in Eastern Europe, but also on the part of the Gaullist movement in France, the hottest consequences of the Cold War were felt largely in the Third World. This term is itself a product of the Cold War. Although it was first coined in France in the late 1940s to refer to a possible Third Estate or Third Way, it soon began to refer to those parts of the world outside the settled spheres of influence of the superpowers, the First and Second Worlds, in which their global conflict would be concentrated (see Chapter 3). This logic was to prove visionary. Geopolitical space was conceptualized in terms of a threefold partition of the world that relied upon the old distinction between traditional and modern and a new one between ideological and free. Actual places became meaningful as they were slotted into these geopolitical categories, regardless of their particular qualities.

The threefold categorization turns on a combination of the essential attributes produced by a cross-classification of the pairs of terms traditional (backward) versus modern and ideological versus free. At a first cut, the modern developed world is distinguished from the traditional underdeveloped world (the Third World). At a second cut, the modern world is divided into two parts: a non-ideological (capitalist) or natural (free) First World and an ideological (socialist) Second World. Of course, this is the dominant US rendition of who is and who is not ideological. In the context of the Cold War only left-wing or socialist politics received this attribution.

The classification has not been contained by its particular origins. Indeed, the concept of the Third World has come to signify resistance to the discursive domination of the superpowers and the possibility of alternative paths to development. For example, theories of development based on the idea of dependency and emanating from such Third World settings as Latin America and Africa reject the argument that the sources of uneven development are to be found within poorer countries themselves; they lie in their external connections to the world economy. The Non-Aligned Movement of such Third World leaders as Tito (Yugoslavia), Nehru (India) and Nasser (Egypt) attempted to subvert the very idea of a simple choice between the United States and the Soviet Union. For dominant political and academic élites on the two defining sides at the height of the Cold War, however, there was little doubt that subordination and imitation were the order of the day. The Third World was that vast geographical zone not yet committed to a particular path to modernity. The success of the superpowers would lie in their relative abilities to recruit candidates for their respective models of political economy from the ranks of the Third World. At the same time, particularly for the Americans, there was the need to counter governments that came to power without a clear commitment to US policies. From Guatemala in 1954 to Grenada in 1983 one or other of these rationales was provided for US foreign intervention. The Soviet Union took a similar line in Hungary, Czechoslovakia and Afghanistan.

In each case the local situation was invariably tied to the larger global context. This was because no particular place had singular attributes, only characteristics that followed from its position in the abstract spaces of the Cold War. It was 'friendly' or 'threatening', 'ours' or 'theirs'. This homogenization of global space made knowing the details of local geography unimportant or trivial. All one needed to know was: whose side are they on – ours or theirs?

Identifying or defining places without a larger classificatory grid is not possible. All maps depend on projections, for example. But local knowledge is possible to a degree that Cold War geopolitics essentially denied. So-called area specialists in government and university departments were frequently frustrated by their lack of influence. Certainly, in the United States the 'big picture experts' like Henry Kissinger, Alexander Haig and Zbigniew Brzezinski were always more important than the area specialists in policy disputes. Local knowledge was 'background' that could be called on when needed. It did not determine the real identity of places. That was done by putting them into the global frame of reference.

Conflicts with apparent local roots were thus read as local manifestations of the superordinate global one. Links to outside powers in the form of supplies of arms or the furnishing of advisers were read as the only causes of local conflicts. States following their own development or foreign policies were seen as candidates for neutralism or 'Finlandization' – a dreaded condition involving trading with the Enemy but without unequivocal commitment to one side or the other. States were not autonomous actors but agents for one side or the other, expected to fulfil the political and economic goals of the superpowers. Formal organizations such as the North Atlantic Treaty Organization (NATO), the International Monetary Fund (IMF) and the World Bank on the US side and the Warsaw Pact and the Council for Mutual Economic Assistance (CMEA) on the Soviet side institutionalized the division. In the various United Nations organizations (such as the Security Council and the General Assembly), created at the end of the Second World War to inaugurate a more pacific world order, the two sides expressed their mutual contempt and hostility. The Third World was given a voice but precious little else in the United Nations Conference on Trade and Development (UNCTAD).

The global spatial division recapitulated on a world scale a tendency strong in both US and Soviet (Russian) cultural history to draw strong lines between the space of the 'Self' and the space of the 'Other'. In both cases, if best known in the US case, political discourse has been shaped around the experience of the *internal frontier* which during national development had separated civilization from savagery, domesticated from wild, good from evil. The entire course of US history is read in many school history texts as a realization of Manifest Destiny: that Americans are a chosen people destined to expand in territory, wealth and influence. Similarly, in the Soviet case, Russians long saw themselves as the bringers of light into the cultural darkness of Siberia and Central Asia. The closeness of the colonial frontier allowed each to think of itself as engaged in something quite different from the nasty imperialism of the real Europeans.

Another incentive to a neat bifurcation of global space came from the possession of large numbers of nuclear weapons by the superpowers. If allies became involved in conflicts there was always the danger of escalation from local/conventional to global/nuclear war as they drew in their respective senior partners. This served to discipline allies since they provided the places where escalation (i.e. nuclear war) would commence. Bipolarity, in the sense of two overwhelming military powers, thus imposed a balance of terror on the world as a whole and froze political boundaries as they were in 1945. But it also restrained the superpowers and, by the 1960s, gave leverage to allies who could exploit the superpowers' fear of mutual destruction once they were able to threaten one another directly with arrays of intercontinental ballistic missiles (ICBMs). Nuclear weapons are not inherently stabilizing even under conditions of bipolarity because there is always a danger of automatic escalation. However, the concentration of nuclear ownership certainly served to highlight the distinctiveness of the two superpowers, whatever their other military deficiencies.

George Kennan, a US official in the Soviet Union at the end of the Second World War, in his famous 'Long Telegram' from Moscow and 'Mr X' article in *Foreign Affairs* in July 1947, argued that the Soviet Union was a totally alien space with which there could not be meaningful compromise. This claim had an important effect on the policy choices that the Truman administration made relative to the Greek Civil War (1947) and the founding of NATO (1949). Little surprise, then, that the image of huge blocs of space without meaningful internal variation became a fundamental part of the Cold War geopolitical imagination.

Three geopolitical concepts played especially important roles in naturalizing these understandings of space and global politics for Americans and others. These were containment, domino effects and hegemonic stability. Containment, first enunciated by Kennan, referred to the military and economic sequestration of the Soviet Union. Russia's historical geography, not simply its cultural difference, was invoked to give this argument scientific respectability. As Kennan (1947, 574) put it,

> The very teachings of Lenin himself require great caution and flexibility in the pursuit of Communist purposes. Again these precepts are fortified by the lessons of Russian history: of centuries of obscure battles between nomadic forces over the stretches of a vast unfortified plain. Here caution, circumspection, flexibility and deception are the valuable qualities, and their value finds natural appreciation in the Russian or oriental mind.

A policy of 'adroit and vigilant application of counter-force at a series of constantly shifting geographical and political points, corresponding to the shifts and maneuvers of Soviet policy' (ibid., 575) was necessary to contain the inherently expansive Soviet Union. Throughout his argument Kennan had recourse to a patriarchal mythology that repeatedly characterized the Soviet Union as a potential seducer and rapist with repressed instincts that could burst forth at any point along its boundary unless there were constant pressure along all its length to keep it contained (Ó Tuathail and Agnew 1992).

Kennan's conception of containment was much more expansive than some later commentators have alleged. But it was at least confined to the margins of the Soviet Union, somewhat after the fashion of Mackinder's 'Heartland' model. Although there is no evidence of a direct link from Mackinder to Kennan, among some post-war American security intellectuals Mackinder's ideas, designed for a very different historical context, took on a prophetic yet scientific role in naturalizing containment as a foreign and military policy.

However, the concept of containment became increasingly expansive. The geopolitical codes of US presidents underwent a series of shifts from the 1940s to the 1980s. Examining presidential State of the Union addresses for their global geographical content, O'Loughlin and Grant (1990, 527) report the following trend:

> In the 1940s and 1950s, emphasis was placed on the perceived threat to the 'Rimland', that zone of containment arranged in a semicircle around the Soviet heartland. In the 1960s, attention to specific conflicts in Cuba and Vietnam was added to the dominant U.S./U.S.S.R. competitive theme. In the 1970s decade of detente, attention to foreign policy was reduced, only to be revived strongly in a regional guise in the late 1970s and early 1980s by Presidents Carter and Reagan. During the 1980s, the regional focus of U.S./U.S.S.R. competition has shifted to the Middle East, Southern Africa and Central America.

The main way in which the scope of containment was enlarged was through the logic of the so-called domino theory or domino effects. This argues that the sooner some potential threat to the global status quo was engaged, wherever it might occur, the less likely was it to produce a spread or contagion effect that could eventually lead all the way back to the United States. In a more sophisticated version the domino theory holds that the credibility of US commitments in more important regions, such as Europe, would be undermined by a failure to protect client regimes in distant corners of the globe. In such circumstances, US resolve to resist any aggression would be opened to doubt and the Great Adversary would be emboldened.

The metaphor of falling dominoes was first used by President Eisenhower in the mid-1950s to describe the consequences of the 'loss' of South Vietnam if the Communist insurgents there prevailed:

> You have a row of dominoes set up, you knock over the first one, and what will happen to the last one is the certainty that it will go over very quickly. So you could have the beginning of a disintegration that would have the most profound influences.
>
> (quoted in Gregory 1978, 275)

But President Truman had used much the same logic in arguing for US intervention in the Greek Civil War in the late 1940s. His metaphor was different:

> Like apples in a barrel infected by one rotten one, the corruption of Greece would infect Iran and all to the East. It would also carry infection to Africa through Asia Minor and Egypt, and to Italy and France, already threatened by the strongest domestic Communist parties in Western Europe. The Soviet Union was playing one of the greatest gambles in history at minimal cost.
>
> (quoted in Gregory 1978, 276)

This logic was invoked again at the outset of the Korean War, repeatedly and explicitly during the Vietnam War, and in Chile and Angola in the 1970s. Finally, President Reagan used it as the centrepiece of his administration's policy towards Central America in the 1980s. The successful overthrow of the Somoza dictatorship in Nicaragua in 1979 led the Reagan administration to support a group of counter-revolutionaries called the Contras committed to removing the Sandinista regime that had recently come to power and that was supported by the Soviet Union and its Caribbean ally Cuba. This support was based primarily on the logic of the domino theory. President Reagan himself speculated openly that if the Sandinista government were not removed, it would destabilize surrounding governments, and very soon Sandinistas would show up at the Mexican border of the United States.

The domino theory is really a chain-reaction or contagion metaphor, whether in the rotten apple or falling dominoes version. It effectively externalizes local conflicts into aspects of the overarching global conflict. It does so by tying US national security into distant places through the possibility of the spread of revolution or the incubus of communism. Political change is thereby converted into a natural epidemiological process that threatens to diffuse like a disease into hitherto uninfected regions – eventually back to the United States itself. The fear of appeasement animating the generation that ran US (and Soviet) foreign policy after the Second World War was expressed in a linkage metaphor that put appeasement (or negotiation) beyond the pale of acceptable discourse.

As US dominance over its part of the globe slipped in the 1970s, a third naturalizing prop was added to Cold War geopolitical discourse. This one involved arguing that interstate cooperation and an optimal world economy needed a benevolent hegemon, such as the United States, for its best possible operation. 'Both Great Britain in the nineteenth century and the United States after the Second World War helped bring about an interdependent and overall peaceful world' (Grunberg 1990, 433). Whatever its empirical merit, which is doubtful, this idea served to 'link the fate of the world with that of the United States'; there was 'no escape from chaos, except through a rejuvenation of U.S. power' (ibid., 447–48).

Rather like a kindly father or a self-sacrificing hero, the United States' destiny was to hold off the cataclysm that would otherwise follow its demise. The justification for its centrality, however, was different. American images of fatherhood were just too ambiguous! Justification was offered in terms of the *systemic* need for a central provider of 'public goods' – lending at last resort, enforcing liberal trading rules, providing a stable unit of monetary account – that would be underprovided without the hegemon because it would be possible for any one state to consume its benefits without paying for them. In an anarchic international system only a benevolent despot can provide enforcement in the common interest. Consequently, the successful working of the world economy required continuing US 'leadership'.

The Cold War geopolitical imagination, therefore, although overtly ideological in the sense of competing commitments to different political–economic models, also offered various naturalized fictions that could be called up for periodic duty. They allowed the Cold War to become a self-fulfilling prophecy and placed it beyond rational consideration. Overall, the articulation of Cold War geopolitics helped to secure and reinforce a set of 'geographical identities ("the West", "the Soviet Union", "the United States"), while serving to discipline domestic social and cultural differences within these spaces' (Ó Tuathail 1993). The polarity between the United States and the Soviet Union was vital to this process. Even though exceptional intellectuals of statecraft, such as President Nixon and his Secretary of State Henry Kissinger in the early 1970s, tried to move US foreign policy from its bipolar mode into a multipolar balance of power configuration, their emphasis on demonstrating 'credibility' wherever US power might be in question led them through the domino analogy right back to bipolarity.

Conclusion

The modern geopolitical imagination, like the configuration of global space it purports to understand, was never set for all time. In this chapter I have surveyed how the modern geopolitical imagination has adapted to and underwritten three eras of modern world politics. The civilizational geopolitics of the early nineteenth century gave way to the naturalized geopolitics of the late nineteenth and early twentieth centuries. After the Second World War an ideological geopolitics was constituted on the basis of the new geopolitical realities of the time. Each of these versions of the modern geopolitical imagination has had its own particular combination of elements. The first was based rather more on a sense of the opposition between European and other spaces that was then taken as justifying European predominance. The second focused particularly on the exclusive claims to territory and empire of competing states whose interests were economic and whose identities were racial. The third rested on the competing models of modernity offered by the United States and the Soviet Union, two states whose histories of expansion and internal organization differ profoundly from those of the classic European Great Powers. But there have also been important continuities. These have been identified in the persisting themes of the geographical projection of backward versus modern flowing through geopolitical discourse from its origins in Renaissance Europe, seeing the world as a single entity and territorial states as the exclusive actors in world politics pursuing strategies of world primacy. The world can now be said to be living in postmodern times because of an increased awareness of the Eurocentric nature of this way of thinking. A key challenge is to survey and critically engage alternative scenarios for the future in as full knowledge as possible of the workings of the modern geopolitical imagination. The next chapter attempts to do this for the world after ideological geopolitics.

Recommended reading

Bassin, M. 1991 Russia between Europe and Asia: the ideological construction of geographical space. *Slavic Review*, 50: 1–17.

Campbell, D. 1992 *Writing Security: United States Foreign Policy and the Politics of Identity.* Minneapolis: University of Minnesota Press.

Dijkink, G. 1996 *National Identity and Geopolitical Visions: Maps of Pride and Pain.* London: Routledge.

Hearn, C. G. 2002 *Tracks in the Sea: Matthew Fontaine Maury and the Mapping of the Oceans.* Camden, ME: International Marine/McGraw-Hill.

Higonnet, M. R. 1994 New cartographies, an introduction, in M. R. Higonnet and J. Templeton (eds) *Reconfigured Spheres: Feminist Explorations of Literary Space.* Amherst: University of Massachusetts Press.

Kern, S. 1983 *The Culture of Time and Space, 1880–1918.* Cambridge, MA: Harvard University Press.

Kristof, L. 1968 The Russian image of Russia, in C. A. Fisher (ed.) *Essays in Political Geography.* London: Methuen.

Pagden, A. 1994 *Lords of All the World: Ideologies of Empire in Spain, Britain and France c.1500–c.1800.* New Haven, CT: Yale University Press.

Perkins, D. 1992 *Is Literary History Possible?* Baltimore: Johns Hopkins University Press.

Pletsch, C. E. 1981 The Three Worlds, or the division of social scientific labor, circa 1950–1975. *Comparative Studies in Society and History*, 23: 565–90.

Sewell, W. H. 1996 Three temporalities: toward an eventful sociology, in T. J. McDonald (ed.) *The Historic Turn in the Human Sciences.* Ann Arbor: University of Michigan Press.

Sharp, J. P. 2000 *Condensing Communism:* 'Reader's Digest' *and American Identity.* Minneapolis: University of Minnesota Press.

Smith, G. 1994 *The Last Years of the Monroe Doctrine, 1945–1993.* New York: Hill and Wang.

Spurr, D. 1993 *The Rhetoric of Empire: Colonial Discourse in Journalism, Travel Writing, and Imperial Administration.* Durham, NC: Duke University Press.

Stephanson, A. 1995 *Manifest Destiny: American Expansion and the Empire of Right.* New York: Hill and Wang.

Taylor, P. J. (ed.) 1993 *The Political Geography of the Twentieth Century.* London: Belhaven Press.

Toulmin, S. 1990 *Cosmopolis: The Hidden Agenda of Modernity.* Chicago: University of Chicago Press.

7 A new age of 'global' geopolitics?

The demise of the Soviet Union as an ideological Other undermined the entire basis of Cold War ideological geopolitics. Particularly in the United States, the protective cocoon and organizing framework that the Cold War provided to political and intellectual life has shrivelled. In its place has come an extreme ontological insecurity, a widespread sense of uncertainty about how to organize world politics in its absence. In some quarters there is even a nostalgia for the 'good old days' when East was East and West was West and never the twain should meet. The US perception of increased insecurity was significantly heightened by the terror attacks of 11 September 2001 on the World Trade Center in New York and the Pentagon in Washington, DC – the first major military assaults on the US mainland since the British army burned the White House during the war of 1812. Though the long-term effects as yet remain unclear, a number of plausible geopolitical scenarios are under construction. This chapter is devoted to describing these.

One point does seem clear: the ideological geopolitics of the Cold War worked because two states pretending to the mantle of modernity confronted one another globally. Neither a militant Islam nor, for example, 'evil drug barons' provide an equally well-defined, competitive and potent substitute, although some commentators and politicians in the United States and Europe do offer demonized portraits of Japan, China and/or the Islamic world that would give this impression. In the absence of either a widely accepted alternative world-view or a convincing replacement for the former Soviet Union the geopolitical imagination must once more be reconstituted. But there are confusing and contradictory signs as to how this will be done. It is difficult to portray a single scenario as to what is in the process of replacing the Cold War as an organizing template for global geopolitics. Three scenarios offer rather distinctive perspectives on what is in course. After briefly describing each of these, I offer a short conclusion about 'global geopolitics'.

Three scenarios

The three most obvious candidates as the basis for a new geopolitics are one that identifies those new practices and representations of a deterritorializing and transnational global economy (often referred to by the term 'globalization'), another that sees the prospect of culture wars between different 'civilizations', and a third that sees the United States as exercising an imperial dominance with little or no military competition. In each case the given scenario can be either welcomed or regretted, seen as either progressive or regressive, but the existence of three such divergent scenarios suggests that as yet no single vision and associated practices has won out over the others. There is evidence for all three. If I tilt towards the first scenario it is because I think that the outcome of the long US–Soviet competition is a world in which transnational networks have become increasingly important in the lives of people just about everywhere. The other two scenarios seem habituated to a world of fixed spaces, in the one case territorial civilizations and in the other state-based imperialism. In neither case is this the world that I see emerging on a daily basis.

The market-access regime and opposition to it

Moreover, in the first scenario there is the possibility of using the plurality of spaces emerging under the influence of the transnational liberalism described in earlier chapters (particularly Chapters 4 and 5) for a kind of anti-geopolitics built upon a commitment to treating places and people as if they counted independently of their global economic and military 'status'. Many elements in the 'anti-globalization' movements that mobilize around the meetings of international organizations and the World Social Forum in Porto Alegre, Brazil, take this approach: creating international coalitions to challenge the depredations of capital mobility but also to welcome a world less compartmentalized territorially and more open to social change. Without such subversion, however, there is likely to be a deepening of the market-access regime of world capitalism and its commitment to a world economy in which capital will be increasingly unconstrained in moving around the globe to exploit differences in rates of return on investment. A world of rich and poor zones, better accounted for in terms of the system of world cities and their hinterlands than by the world political map, will progressively challenge the contemporary hierarchy of states (see Figure 7.1). Whichever trend becomes dominant – the anti-geopolitics of protest or the increased scope of the market-access regime – there is a limited long-term likelihood of returning to the contours of world politics as they were defined during the previous three ages of geopolitics.

From this point of view, therefore, whatever happens will be something new in which the old principles will be of diminished relevance. In the interim, however, they will not readily disappear. Ideas about relative decline, state-territorial competition, plotting state strategy on a global basis, and thinking of development in terms of backwardness and modernity are so deep-seated as to defy easy supersession. The vocabulary of state-centred geopolitics is still underwritten materially by pressures from social groups reacting to increased economic globalization by reviving ethnic and local identities or attempting to resurrect state powers. Strong national identities die hard. States still provide the main opportunity structures for most forms of political activity, even as contra-governmental forces increase in number and scope. Governmentality (territorialized regimes of governmental authority) and movements/trends counter to this now exist in an unstable tension. Even if transnational businesses have increased their powers, states still provide the regulatory environments in which these businesses must operate. Opponents of market-access policies are still largely constrained to work through existing institutional channels, most of which are still state organized. The US government remains the single most important sponsor and source of support for globalization of trade, finance and production.

The events of 11 September 2001 and subsequent economic recession in the United States and around the world have put something of a damper on the rather more far-fetched globalization scenarios. The US government has acted to reinstate more effec-

Figure 7.1 The world during (A) and after (B) the Cold War.

Source: Agnew and Corbridge (1995)

tive border controls after allowing for a loosening up in previous years to encourage the cross-border movement of goods, capital and, more problematically, of people. As part of the 'war on terrorism', the US government has also targeted states held to support terrorism or pose a likely military threat to the United States or its allies. Rather than seeing global terrorism as a qualitatively new phenomenon, therefore, there has been a tendency to squeeze it back into a state-centred mould. The US government has also become increasingly protectionist in relation to agriculture and steel imports, using the attacks as a pretext to bolster the territorial economy of the United States in the face of foreign competition.

Whether this reinstatement of boundaries signifies a wholesale return to a heavily bordered world of states, however, is open to considerable doubt. The growth of the world economy has become based upon a wide range of inter-state exchanges. The US economy, for example, depends crucially on the import of capital to make up for the lack of domestic saving. The recent explosive growth of China's economy is premised on ready access for its manufactured goods to US and other foreign markets. Capital markets are increasingly integrated across national boundaries. Small states now have the ability to leverage themselves into the world economy on advantageous terms. This has resulted in the top ten of the international league table of highest income per capita countries being dominated by small and intermediate-size countries. As the population size/economic growth equation has been rewritten, the likelihood of military conflict and territorial expansion resulting in hefty economic returns has been much reduced. The world of the market-access regime is radically different from the worlds that preceded it, hence the kind of politics that is required to challenge its depredations is likely to be different from the politics of the past. 'Capturing the state', for example, as Leninists and social democrats have long advocated as the best strategy for embarking on schemes of radical social change, looks like a strategy whose time has passed. Whether or not this requires the substitution of placeless global networks of a 'multi-tude' of opponents (as in Hardt and Negri's (2000) *Empire*) is another thing entirely. Indeed, this is likely to be an exercise in frustration as long as the world remains terri-torially divided and political identities are rooted in place. What seems more fruitful in fact is to represent the worst excesses of the market-access regime, particularly the fiscal programmes championed by the IMF and World Bank, as destructive of people's loyal-ties to place and thereby to ground political opposition not in the nebulous ether of 'being against' but in relation to the realities of people's everyday lives.

The era of globalization since the 1970s has had undeniable effects, not entirely nega-tive. For one thing, states such as China and India have seen remarkable rates of economic growth that perhaps augur the beginning of a long-term shift in the distribu-tion of global income. For another, the historical record suggests that the world has become a relatively less violent place, in the sense of major inter-state wars at least, than it was in the twentieth century. Finally, women and many other groups 'naturally' excluded from politics in times past have forced themselves into the picture. The story of globalization, therefore, has very positive aspects that many of its critics fail to note, often retreating into nostalgia for a state-dominated past or waxing utopian about tribal or feudal modes of production as plausible alternatives to global capitalism.

The changes associated with globalization have been enabled by a host of techno-logical, economic and cultural changes – from just-in-time warehousing, shipping containerization, the fax machine, the Internet, and easy round-the-world air travel through offshore production and integrated world commodity, currency and stock markets to worldwide consumer tastes, the massive spread of diasporic communities, and the mass availability of 'exotic' vacations. But perhaps these changes augur a more profound shift than simply the redistribution of global incomes, diminished inter-state war and the redistribution of power. Such would be the view of those for whom a 'post-modern geopolitics' is in the offing. From this perspective, globalization 'is the key trope in tying together neo-liberal capitalist rationality, informational technics, mass consumption culture, and the integrated world markets of a postmodern geopolitics' (Luke 2003, 228). The consequence is that 'What were once discrete "solid state" circuitries for geopolitical power in closed hierarchical systems must face open-sourced architectures of power in which capital and authority work at nodes in networks' (ibid., 229–30). How can the modern geopolitical imagination possibly survive these pressures over the long term?

The clash of civilizations

The 'culture wars' scenario has become particularly popular with those looking to reconstitute the ideological geopolitics of the Cold War on a multipolar basis. Cold War geopolitics rested on an essential opposition drawn between what were supposed to be two completely different types of society/culture. In reality, of course, cultural differences are always relative ones and there is much in common between cultures that when isolated one from the other appear more particularistic than they really are. Both 'Islamic' and 'Confucian–Asian' cultures, for example, combine influences from other parts of the world (above all Europe) and have changed profoundly over the years. The image of fixed, isolated cultures should be rejected for the ideological imposition that it is. Yet an increasing number of writers and intellectuals of statecraft are placing stress on the importance of cultural values and institutions in the geopolitical confusion left in the wake of the Cold War.

To the American political scientist Samuel Huntington (1993), for example, the most influential exponent of the 'clash of civilizations', future wars will occur between the nations and groups of 'different civilizations' – Western, Confucian, Japanese, Islamic, Hindu, Orthodox Christian and Latin American, perhaps also African and Buddhist (see Figure 7.2). The fault lines between these cultures will define the geopolitical battle-lines

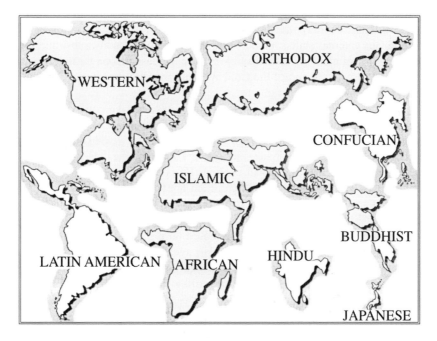

Figure 7.2 Huntington's civilizations (1993). Each civilization is identified as entirely discrete to convey visually the relative size and range of the civilizations as a set.

Source: Redrawn from Huntington (1993)

of the future. 'Culture and cultural identities . . . are shaping the patterns of cohesion, disintegration and conflict in the post-cold war world. . . . Global politics is being reconfigured along cultural lines' (ibid., 23).

Huntington's maps seem to serve a technical, heuristic purpose. They simply describe the 'reality' of a world divided along cultural fault lines and signal where in the world conflicts will erupt. This is how the modern geopolitical imagination has always imagined maps to be. What is new with Huntington is simply that territorially discrete 'civilizations' substitute for territorial states that now become the 'agents' of the civilizations of which they are part. Maps, however, as Chapters 2 and 3 suggest, reflect their makers as much as or more than the world they purport to describe. In this case, to produce the discrete civilizations Huntington must 'freeze-frame culture' into fixed and timeless mentalities that reflect his own vision more than the ambiguous and shifting spatialities of the world that many of us know (Shapiro 1999, 6).

The problem with Huntington's scenario is twofold. First, who identifies with such broad-scale 'cultures'? Certainly the 'Western' category is problematic, as contemporary attempts at creating a common sense of 'Europeanness' (in the European Union) founder on the shoals of revived and invented national and ethnic identities. Increasing divergence between Europe and the United States across a range of policy issues also suggests that the 'Western' category is no longer as self-evidently homogeneous as it once seemed. Second, globalization undermines cultural closure. The increasing flows of information, goods, people and capital around certain parts of the world (particularly between Europe, North America and East Asia) not only cause potential friction between cultures, but also tie cultures together, and increase tensions within culture areas as different social groups and individuals make different judgements about this or that external influence. Except on a super-organic view that sees them as external to populations, cultures are never set in stone; they adapt over time in response to external pressures and internal shifts. 'Culture' cannot, therefore, readily substitute for the all-embracing role of ideology in the Cold War. The increased scope for independent action by more states with the end of the Cold War and the pressures imposed on all of them by globalization work against it.

This said, the 'clash of civilizations' scenario has received something of a vindication in the aftermath of the terror attacks of 11 September 2001. The apparent agent of the attacks, the Islamic fundamentalist network *al-Qaeda*, is led by the Saudi millionaire Osama bin Laden. Bin Laden accepts the logic of the 'clash'. He is a prophet and organizer of inter-civilizational conflict. He is the modern Arab geopolitician par excellence. Actively pursued by US forces but not found (as of late 2002), he is not an operative as much as a theoretician of what divides the Islamic and Western 'civilizations' (Plate 7.1). Mesmerizing and charismatic in his effect on those who come into his company, he has a clear world-view based on a specific, if peculiar and heretical, reading of Islam's holy book, the *Qu'ran*. His agents recruit adherents in a manner akin to that used by the Reverend Moon and his Unification Church (and other cults): approaching lonely, vulnerable young men affiliated with mosques and Islamic centres throughout the Arab diaspora, particularly in Germany, England and North America, and university students in Egypt, Algeria and Saudi Arabia, using their alienation from the Western world to turn them into his zealots. Exploiting the open borders, rapid transportation and telecommunications of a globalizing world, he is bent on establishing an alternative geopolitical world to the US-dominated one now in the offing. A product of globalization, he is anything but a 'traditional' figure, however ancient he might claim

"*You really think this will work?*"

Plate 7.1 'You really think this will work?' This cartoon identifies one of the primary difficulties of the US 'war on terrorism': finding Osama bin Laden, the leader of the *al-Qaeda* terrorist network allegedly responsible for the attacks of 11 September 2001.

Source: © The New Yorker Collection 2002, Mick Stevens from cartoonbank.com. All rights reserved

the tradition is that he has been inventing. He offers a mirror-image security mapping of the world to that offered by Huntington and other prophets of the 'clash of civilizations'. Like the Western pundits, bin Laden collapses ontology into geography – the key move of the modern geopolitical imagination. Vice and virtue have geographical addresses. Like the pundits, he fails to see the arbitrariness of the 'civilizational codes' that mark the world. Unlike them, he sees the Islamic world as the superior one to be protected from the contamination of the West. Like them, he despises multiculturalism, gender equality and tolerance, associating these social ideals with degradation and weakness. Like them, and in departure from orthodox geopolitical thinking, he rejects the world of states as the future model of world politics. His priority lies in securing the home-space for an Islamic civilization defined against a West that has dominated the Arab world for far too long.

Bin Laden's problems will be threefold, whatever the initial and continuing success of his terrorist activities. The first is that the version of Islam he represents is a peculiarly puritanical and violent one. As commentators on Huntington have noted, 'contemporary Islam harbors a complex and pluralistic set of relationships between religious and political institutions' (Shapiro 1999, 4). The particularly anomic conditions of Afghanistan following the Soviet invasion of 1979–89, subsequent civil war, and drought may have proved favourable to the spread of bin Laden's view of the world among the largely Pushtun supporters of the Taliban regime. Alienated young men from the Arab world in the cheap hotels and student hostels of Western Europe may well continue to prove ready recruits to his ranks. But the nationalist aspirations of the Palestinians, the divergent interpretations of Islam between Sunni and Shia branches, the divisions between Arabs and other Muslims, the tolerant Islam characteristic of most American Muslims, and the deep-seated attractions of consumerism and/or democracy all over the Arab world will work against him. Opinion polls show that most Muslim and Arab respondents claim to share many of the same values as Americans and other 'Westerners', approve of democratic ideals, and, though disapproving of US foreign policy in the Middle East, would like 'better relations' for their countries with the United States (*The Economist*, 19 October 2002, 43).

The second problem for bin Laden and *al-Qaeda* is that the terrorist acts of 11 September 2001 have helped to resecure US and, more broadly, Western identities that had begun to erode in the aftermath of the Cold War. During the Cold War the Soviet Union represented the Other against which American identity was defined. In this light, American identity was underwritten by US foreign policy. In the 1990s there was considerable nostalgia in the United States, particularly among intellectuals of statecraft and in the military–industrial complex, for the world that had provided them with a sense of purpose and with fat paychecks. In the absence of an overwhelming threat or danger, such as the former Soviet Union and its missile forces, US foreign policy had lost both purpose and coherence. This removed one of the main props upon which a distinctive American political identity had rested for fifty years. Bin Laden has undoubtedly helped to re-establish a strong sense of an external danger that cannot be ignored and that can reorient US foreign policy. An open-ended 'war on terrorism' is the result, in which the might of the United States is directed towards trying to secure the US 'homeland' from both domestic enemies and foreign foes. Osama bin Laden is the new enemy number one around which US foreign policy can be reorganized and a reinvigorated national identity can be redefined.

The third problem is one that can and must be created for bin Laden and his supporters. It is to challenge the plausibility of the story of a singular 'West', the moral geography, if you will, that bin Laden thrives on: the secular, consumerist, spiritually bankrupt, amoral and exclusionary world that values nothing other than the 'bottom line' and continued economic growth at any price. In the shadow cast by the worst terrorist act visited on US soil, Americans can begin to ask what they have wrought in the world and how they should change to make the whole world a better place. Of course, Americans could just continue to be themselves, imposing military and economic hegemony to keep the world in line. But they, and others, could start to challenge the geopolitical abstractions that turn places into geopolitical commodities, people into pawns of superpowers and terror networks that aspire to similar power, whether it be to produce the rubble of Kabul or the now-downed towers of the World Trade Center.

From state-centrism to US unipolarity

A third scenario is sceptical that anything much has changed about the role of states in particular and the nature of geopolitics in general. From this viewpoint, all that has changed is that one state, the United States, has become predominant globally. Prepared to adopt a hands-off attitude before 11 September 2001, proposing a national missile shield, opposing international agreements, and generally adopting a hostile attitude towards the idea of an 'international community', the US government then decided to go on the offensive. Shifting from a defensive to an offensive 'realism' means making a bid for empire even if posed in terms of a perpetual 'war on terrorism' and the states that are held to support terrorism (such as Iraq and its leader, Saddam Hussein). In this scenario the twenty-first century bodes well to become the real 'American Century'.

The basic premise of this scenario is that there are no significant alternatives to the exercise of US power. This power has three main aspects to it, weighted in different degrees by different commentators and practitioners. The dilemma is that these three are not necessarily compatible and, indeed, show signs of being inherently contradictory. The first is military. In this category the United States simply has no peer (Figure 7.3). Contrary to predictions in the 1980s of US decline, with the demise of the Soviet Union the United States has established itself as a global military hyperpower. The United States has a seemingly unassailable technological edge in war-fighting capabilities. At one and the same time this tempts US governments into adventurous interventions around the world yet also makes the US 'homeland' vulnerable to asymmetric warfare: the use of terrorism either with or without weapons of mass destruction. If the 2002 declaration of war on Iraq fits the first category, then the terrorist assaults of 2001 fit the second one. The military prowess and the plans to use it reflect two fears: the growth of global terrorism directed at US targets at home and abroad and the rise of potential global challengers, particularly China. These have become the major priorities since the advent of the G. W. Bush administration in 2001. If the Clinton administrations of the 1990s had undertaken the task of integrating the other major states of the world into a rule-based transnational–liberal order, the Bush administration has become largely committed to an ambitious reinstatement of offensive realism:

> preventing by every possible means the emergence of any serious rival or combination of rivals to the US, anywhere in the world, and to opposing not just any rival would-be world hegemon, but even the ability of other states to play the role of great power within their own regions.
>
> (Lieven 2002, 9)

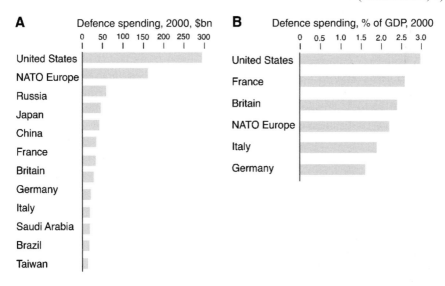

Figure 7.3 Defence spending by country, absolute (A) and relative (B), 2000.

Source: Author from various sources

Not surprisingly, this approach has proved popular with nationalist groupings in the United States and in foreign countries, such as Israel, that perceive benefits for themselves from a more aggressive US foreign policy. 'Redrawing the map' of the Middle East by overthrowing Saddam Hussein in Iraq and threatening the current regimes in Iran and Syria, for example, serves to guarantee both Israeli military superiority in the region and the long-term supply of oil to a voracious US economy. What is more surprising is that the 'return' of an overt state-centrism in the form of a prospective US quest for global imperium as the centrepiece of world politics finds support from elements on the political left (for example, Gowan 2002), both for its rejection of the 'cant' and 'hypocrisy' of the Clinton talk of the 'world community', etc., and for the reinstatement of the state as a central actor and target for political change. Nostalgia for old certainties dies hard across the political spectrum.

But there are countervailing forces. As many commentators have pointed out, including, for example, Anatol Lieven (2002) and John Ikenberry (2002), the secret of the United States' previous success as a hegemonic power has rested on its ability to persuade others to follow its lead. Moreover, 'unilateral world domination through absolute military superiority' is likely to prove unattractive to large sections of US public opinion and those sectors of US capital engaged in transnational business (Lieven 2002). This is so not least because the economic basis of US power presses in a contradictory direction.

The second dimension is indeed that of US economic power. As the dominant force in the world economy since the Second World War, the United States has worked to open up the world economy to the relatively free flow of capital and trade so as to benefit its national economy and the capitalists who dominate it. This long-term economic basis to US power, however, might seem to work against the neo-imperial model implicit in the militarization of US foreign policy after 2001. But, paradoxically, this is often construed as yet another channel for the unilateral exercise of US power! Undoubtedly, favourable access to or control over key resources, such as oil, does work in this direction but, in fact, much of US economic power has depended precisely on avoiding direct control over distant places. The success of the United States as a Great Power has involved crushing or undermining statist empires of one sort or another. If in the 1940s they were Nazi Germany and the Japanese Empire, from the 1940s until the 1980s it was the Soviet Union. Today its economic policies are eroding the state-directed neo-mercantilism of Japan and South Korea. China has already moved down the road towards a more freewheeling capitalist economy stimulated in large part by the openness of the US economy to Chinese exports. The flip side of US containment of the Soviet Union during the Cold War was an internationalism committed to *spreading*, not restricting, economic development so as to revive and stabilize the US economy. Though this was frequently at odds with the military side of the Cold War, and specifically in US reactions to anti-imperial revolutions in Korea, China and Vietnam, it is now implicit in the Deng Xiaoping adage that 'to get rich is glorious'. A US-made world economy, therefore, has provided the basis for the globalization that now knits major parts of the world together. Unfortunately, the US economy has not always been its 'natural' beneficiary. Indeed, in the 1970s and 1980s the United States suffered economically from the very policies that its governments had earlier set in train. During the 1990s the Clinton administration embarked on an aggressive economic policy to do something about this by promoting US-manufactured exports, opening up targeted economies to US service exports, and using the US dollar to manipulate the world economy to US advantage.

Whether these sorts of actions will invariably produce a sustainable benefit for the United States is still an open question. The very success of US opening up of the world economy now makes the United States itself vulnerable to external shocks. What is less doubtful is that an aggressive US imperialism will impose economic costs that the US economy is increasingly ill-equipped to cover. Overthrowing regimes and reconstructing countries, as proposed for Iraq in 2002, is an expensive business for a country such as the United States, which, although the world's biggest national economy, also has the world's largest balance of payments deficit and a massive and increasing government debt (Nordhaus 2002).

Finally, US governmental leaders are the only actors free to choose the balance of force and consent in the hegemonic calculus at the heart of world politics. But the military and the economic–cultural dimensions of the US presence in world politics are potentially very much at cross-purposes. During the Cold War they tended to work together because they were directed at different places. Military power was mainly directed towards curbing the power and influence of the Soviet Union. Political–economic clout was focused, rather, upon rebuilding the war-shattered economies of Europe and Japan and tying them into a Pax Americana. With the end of the Cold War, US governments are faced with the challenge of how to balance between coercion and consent. In the absence of a 'common threat' this has proven very difficult to do, hence the flip-flop between the Clinton and G. W. Bush administrations. Temporarily at least, following 11 September 2001, the military theme is in the ascendant, fuelled by popular American nationalism, the new electronic weapons that only the US military has available to it, fears about access to oil supplies, and the appeal of the open-ended 'war on terrorism'. But this recipe has gone down particularly poorly in Europe because the European political classes and their populations have become habituated to a different relationship with the United States. The hegemony that the United States has previously exercised had a mix of coercion and consent implicit in it, especially in relation to Europe. Tilting too much away from this may undermine the US global political position precisely at the time when its military power is at its zenith. Unlike previous global hegemons, if the United States chooses empire it may succeed only in eroding the very world it has long claimed to be creating.

Conclusion

What to make of these different scenarios? It seems to me that there is an inherent tension between the globalizing world that the United States has sponsored (and that now has also spawned global terror networks and international drug traffickers) and the US role as a neo-imperial force in world politics. If there is as yet no obvious challenger to the United States as a global hegemon, there is the increasing possibility of a world in which US hegemony is increasingly challenged if it takes a unilateral and coercive basis. This may well be an auspicious moment, therefore, to ask whether the modern geopolitical imagination, which lies at the root of all of these scenarios to one degree or another, is not itself overdue for retirement from the global scene. The attraction of globalization lies partly in the possibilities it offers for a world in which the cry 'Hannibal ad portas' ('Hannibal is at the door', the Roman version of the later 'the Russians are coming') will no longer be heard. As yet, however, we can have little confidence that this will indeed be the case. The world is still all too much in the hold of another adage that has prevailed for centuries but that has never yet worked to produce what it promises: *si vis pacem, para bellum* ('If you want peace, prepare for war').

Recommended reading

Agnew, J. and Corbridge, S. 1995 *Mastering Space: Hegemony, Territory, and International Political Economy*. London: Routledge, chapters 7 and 8.

Arrighi, G. and Silver, B. (eds) 1999 *Chaos and Governance in the Modern World-System*. Minneapolis: University of Minnesota Press.

Brzezinski, Z. 1997 *The Grand Chessboard: American Primacy and Its Geostrategic Imperatives*. New York: Basic Books.

Campbell, D. 1992 *Writing Security: United States Foreign Policy and the Politics of Identity*. Minneapolis: University of Minnesota Press.

Caraley, D. J. (ed.) 2002 *September 11, Terrorist Attacks, and U.S. Foreign Policy*. New York: Academy of Political Science.

Chace, J. 2002 Imperial America and the common interest. *World Policy Journal*, 19: 1–9.

Cumings, B. 1999 Still the American Century. *Review of International Studies*, 25: 271–99.

Desai, M. 2002 *Marx's Revenge: The Resurgence of Capitalism and the Death of Statist Socialism*. London: Verso.

Gowan, P. 1999 *The Global Gamble: Washington's Faustian Bid for Global Dominance*. London: Verso.

Gowan, P. 2002 A calculus of power. *New Left Review*, 16: 47–67.

Hardt, M. and Negri, A. 2000 *Empire*. Cambridge, MA: Harvard University Press.

Huntington, S. P. 1993 The clash of civilizations? *Foreign Affairs*, 72: 22–49.

Ikenberry, G. J. 2002 America's imperial ambitions. *Foreign Affairs*, 81: 44–60.

Lieven, A. 2002 The push for war. *London Review of Books*, 3 October: 8–11.

Luke, T. W. 2003 Postmodern geopolitics: the case of the 9.11 terrorist attacks. In J. Agnew, K. Mitchell and G. Toal (eds) *A Companion to Political Geography*. Oxford: Blackwell.

Mertes, T. 2002 Grass-roots globalism. *New Left Review*, 17: 101–10.

Nordhaus, W. D. 2002 Iraq: the economic consequences of war. *New York Review of Books*, 49, 5 December: 9–12.

Sader, E. 2002 Beyond civil society: the Left after Porto Alegre. *New Left Review*, 17: 87–99.

Shapiro, M. J. 1999 Samuel Huntington's moral geography. *Theory and Event*, 2/4: 1–11.

Soros, G. 1998–99 Capitalism's last chance? *Foreign Policy*, 113: 66.

Tolan, S. 2002 Beyond regime change. The administration doesn't simply want to oust Saddam Hussein. It wants to redraw the map of the Middle East. *Los Angeles Times*, 1 December: M1–6.

Wade, R. 1998–99 The fight over capital flows. *Foreign Policy*, 113: 41–54.

8 Conclusion

The purpose of this book is to trace analytically the beginnings and development of the modern geopolitical imagination by exposing its component parts and showing how these came together under different material conditions and as a result of contestation to produce a succession of different geopolitical epochs. In this Conclusion I restate the premises underlying the book, address a number of possible criticisms that might be directed at it, and then lay out some ideas for an alternative way of considering world political geography.

Revisiting theoretical premises

Two theoretical premises inform the perspective of the book. The first is that in the realm of geopolitics the intellectual and the political are not separable. Geopolitics has served statecraft, usually that of particular states. Yet it has claimed to do so under the sign of objectivity – most typically in the form of the view from nowhere. This contradiction is more than a little problematic. I take the position that it undermines the entire approach. Rather like the explorers of Australia discussed in Chapter 3, students and practitioners of world politics have found what they wanted to find because they knew all along. What was known kept the geopolitical treadmill turning indefinitely. They could never be surprised. They could never see that there was a native point of view that might be worth taking seriously. There was no possibility of searching for communalities and common understandings between us and them. Just as Caesar said and did when he arrived in Britain: We came, we knew, we conquered.

The second premise is that the making of the modern geopolitical imagination has not been a once-only thing. Though the basic building-blocks or principles have remained more or less the same, the end product has gone through a number of important transformations as the political world it has purported both to reveal and to reflect has changed. An imagination or ideas do not exist 'out there' or simply in texts and documents; they are implicit in practices or social action. To survive and prosper they must be passed on from generation to generation as a form of common sense or guide to action and must adapt successfully to challenges and changed historical contexts.

Engaging with criticisms

These premises and the arguments I develop from them are somewhat different from those of other recent commentaries on political geography, world politics and geopolitics. Of first importance, a modern geopolitical imagination is identified as preceding the first use of the word 'geopolitics' itself. I see no problem with this. Only an extreme linguistic nominalism that endows words with a singular causal power would have it otherwise. I present a historical argument to the effect that geographical understandings of a certain type have informed the actions of political élites and that political geography (and other fields) have taken these understandings as the basis for their understanding. Although textual analysis of writers explicitly using the term 'geopolitics' is certainly not excluded by this approach, remaining solely concerned with it strikes me as limiting. The role of the modern geopolitical imagination in world politics is best shown by its practical consequences over a long period of time rather than by examining the claims of its disciplinary exponents. A challenge for some contemporary writers is to decide whether they are primarily interested in the history of geographic thought or in the history of the modern geopolitical imagination, or at least to be clear about which is which.

The approach that I take to understanding the making of the modern geopolitical imagination is a historicist one; it does not presume a set of fixed beliefs, knowledge, attitudes and practices that remain essentially unchanged down the centuries. To the contrary, it presumes change in the content of the geopolitical imagination as a result of changing times. Changing technological, economic, social and political conditions are particularly important. This too is a controversial position. The tendency is to cite some feature from the distant or recent past – say, understandings of the meanings of racial categories – and then project these forward or backward in time as if they were structural features of thought rather than ones with distinctive historical trajectories. In each chapter, and above all in Chapter 6, I have tried to indicate how dependent on historical context the content and deployment of the modern geopolitical imagination have been. Whatever the deficiencies of periodization used in Chapter 6, the appending of the term 'modern' to geopolitical imagination suggests the strong sense of historical dependence that I believe any accounting of the geographical understandings at work in world politics must presume. This is readily accepted when considering the recent breakdown of Cold War geopolitics, as signalled by such phrases as 'rethinking geopolitics', but it seems harder to accept when examining the past. The fear of criticism of whatever periodization one adopts seems to provoke hostility to thinking historically at all.

Beyond geopolitics

A theme running through all the chapters is that present conditions are probably not propitious for the continuation of the geopolitical imagination as we have known it in the past, though there are continuing attempts at breathing new life into it (see, for example, Brzezinski 1997). Indeed, the fact that it has lost much of its intellectual magic, if not yet all its political appeal, suggests that the possibility of rethinking geopolitics also depends on changed historical conditions. It is certainly an appropriate time to rethink political geography's historical dependence on the geopolitical imagination. Again, other commentators would differ. In my view, contemporary conditions suggest the need to abandon the a-priori commitment to the global and national-state geographical scales as somehow exhausting the sum total of causal influences on world politics. This means rejecting both the status quo and the search for a new geopolitical imagination that simply builds on the old foundations. Rather, what is needed is a *geographical* imagination that takes places seriously as the settings for human life and tries to understand world politics in terms of its impacts on the material welfare and identities of people in different places. This involves addressing questions of national and other identities under conditions of massive population movement and diaspora; growing global inequalities; the increased role of regions and localities in regulating economic growth; the rise of supranationalism on world-regional and international scales; and, above all, the growing impacts of globalization of production and finance on states and localities around the world.

In the modern geopolitical imagination political power has been narrowly defined as the ability to make others do something you (person, state) desire and, at least from the nineteenth century, it has been increasingly associated with territorial states. The spatiality of power, however, need not be invariably reduced to state territoriality. At least four models of the spatiality of power can be identified, each of which has dominated in different epochs of geopolitical order. Each of these models is closely associated with a particular set of political–economic and technological conditions. The dominant spatiality of power has changed, therefore, as material conditions and dominant modes of understanding of them have changed. This approach to the historicity of spatiality implies that both material forces and intellectual perspectives or representations interact to produce the spatiality of power predominant within a given historical era. But each also has a synchronic validity in the sense that political power in any epoch can never be totally reduced to any one of them. Rather, in a sense equivalent to Karl Polanyi's (1944) justly famous discussion of market society in terms of the emergence of market exchange at the expense of reciprocity and redistribution as principles of economic integration, as one model comes to predominate, others are not so much eclipsed as placed into subordinate or emergent roles. The models offer, then, a way not only of historicizing political power, but also of accounting for the complexity of the spatiality of power during any particular historical epoch (Figure 8.1).

The first model is that of an 'ensemble of worlds'. In this model, human groups live in separate cultural areas or civilizations with limited communication and interaction between them. The nodes here are 'civilizations'. Each area has a sense of a profound difference beyond its own boundaries without any conception of the particular character of the others. Communal forms of social construction take place within a territorial setting of permanent settlement with flows of migrants and seasonal movements but with fuzzy exterior boundaries. Time is cyclical or seasonal, with dynasties and seasons replacing one another in natural sequence. Political power is largely internally oriented and directed towards dynastic maintenance and internal order. Its spatiality rests on a strongly physical conception of space as distance to be overcome or circulation to be managed.

The second model is that of a 'field of forces'. The nodes here are states. This is the geopolitical model of states as rigidly defined territorial units in which each state can gain power only at the expense of others and each has total control over its own territory. It is akin to a field of forces in mechanics in which the states exert force on one another, and the outcome of the mechanical contest depends on the populations and resources each can bring to bear. Success also depends on creating blocs of allies or clients and identifying spatial points of weakness and vulnerability in the situation of one's adversaries. All the attributes of politics, such as rights, representation, legitimacy and citizenship, are restricted to the territories of individual states. The presumption is that the realm of geopolitics is beyond such concerns. Force and the potential use of force rule supreme

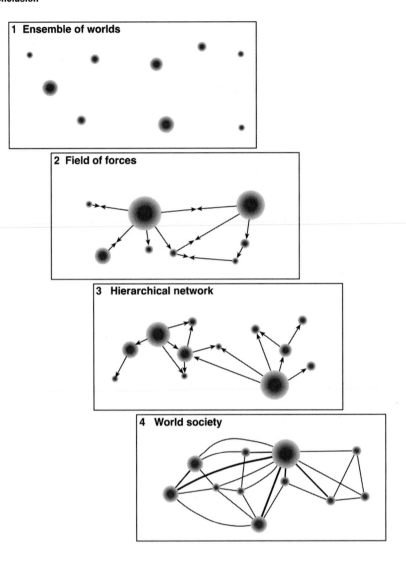

Figure 8.1 Spatialities of power.

Source: Redrawn from Durand *et al.* (1992, 18)

beyond state boundaries. Time is ordered on a rational global basis so the trains can run on time, workers can get to work on time and military forces can coordinate their activities. The dominant spatiality, therefore, is that of state-territoriality, in which political boundaries provide the containers for the majority of social, economic and political activities.

Third on the list of models is that of the 'hierarchical network'. The nodes here are city-regions. The spatial structure in which they are embedded is a world economy of geographical cores, peripheries and semi-peripheries linked together by flows of goods, people and investment. Transactions based largely on market exchange produce patterns of uneven development as flows move wealth through networks of trade and communication producing regional concentrations of relative wealth and poverty. At the local scale, particularly that of urban centres, hinterlands are drawn into connection with a larger world that has become progressively more planetary in geographical scope over the past five hundred years. Political power is a function of whereabouts in the hierarchy of sites from global centres to rural peripheries a place is located. Time is that organized by the geographical scope and temporal rhythm of financial and economic transactions. The spatiality is that of spatial networks joining together a hierarchy of nodes and areas which are connected by flows of people, goods, capital and information. Today, such networks are particularly important in linking together the city-regions that constitute the nodes around which the global economy is increasingly organized.

The fourth, and final, model is that of the 'integrated world society'. The nodes here are social groupings. This model conforms to the humanistic ideal of a world in which cultural community, political identity and economic integration are all structured at a global scale. But it also reflects the increased perception of common global problems (such as environmental ones) that do not respect state borders, the futility of armed inter-state conflict in the presence of nuclear weapons and the advantages of defence over offence in modern warfare, and the growth of an international 'public opinion'. This model privileges global-scale communication based on networks among multiple actors that are relatively unhierarchical and more or less dense depending upon the volition of actors themselves. The sprout-like character of these connections leads some to see them as (in a term popularized by the philosopher Gilles Deleuze) somewhat like the rhizomes of certain plants that spread by casting out shoots in multiple but unpredictable directions. Time and space are both defined by the spontaneous and reciprocal timing and spacing of human activities. Real and virtual space become indistinguishable. This model obviously has a strong utopian element to it but does also reflect some emergent properties of the more interconnected world that is presently in construction.

In the contemporary world there is evidence for the effective co-presence of each of these models, with the former territorial models somewhat in eclipse and the latter network models somewhat in resurgence after a one-hundred-year period in which the field of forces model was pre-eminent (if hardly exclusive). If the trend towards regional separatism portends a fragmentation that can reinforce the field of forces model as new states emerge, then economic globalization and global cultural unification work to reinforce the hierarchical network and integrated world society models. At the same time, movement towards political–economic unification (as in the European Union) and the development of cultural movements with a strong territorial element (as with those Islamic fundamentalist movements looking towards a pan-Islamic territorial polity) tend to create pressures for the reassertion of an ensemble of worlds.

Historically, however, there has been a movement from one to another model as a hegemonic or directing element. In this spirit a theoretical scheme can be suggested in which, first of all, the ensemble of worlds model slowly gave way to the field of forces model around AD 1500 (Figure 8.2). As this was establishing its dominance, the hierarchical network began its rise in and around the framework provided by the state system. Under European colonialism, the part of the world in which states recognized one another as legitimate actors was divorced from the regions in which such status was denied. Since 1945 the hierarchical network model has become more and more central to the distribution of political power. With the end of the Cold War, which had produced

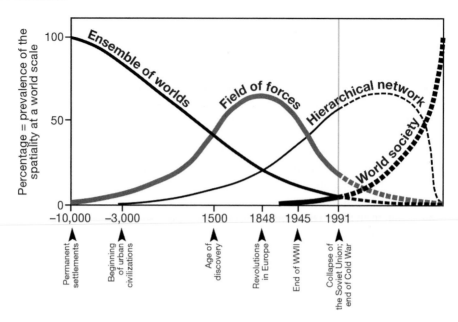

Figure 8.2 Diachronics of spatialities of power.

Source: Redrawn from Durand *et al.* (1992, 18)

an important reinstatement of the field of forces model, the hierarchical network model is in the ascendancy, with signs of the beginning of a trend towards an integrated world-society model. But this is as yet very much in its infancy.

This framework is, of course, only suggestive of long-term tendencies. What it does provide is a sense of the historical spatiality of political power, associated in different epochs with different dominant modes of spatiality. From this point of view, the state-territorial scale is only one among a number of geographical scales that are implicated in the organization and expression of political power. In each of the four models, different relationships between geographical scales are highlighted: from the world-regional and local scales in the ensemble of worlds model, through the state-territorial and global scales of the field of forces model, to the complex core–periphery and global–local connections of the hierarchical network and integrated world-society models. What is clear is that the state-territorial conception of power is not a transcendental feature of modern human history but, rather, a historically contingent feature of the relationship between geographical scales in the definition and concentration of political practices.

A political and intellectual agenda based around thinking about power in terms of different geographical models will be subverted, however, if it is seen only as 'corrective' to the modern geopolitical imagination. After too many generations of state-centred and global geographical determinism, and following the centennial of the first use of the word 'geopolitics', political geographers and others must finally choose whether to be agents of an imagination that has imposed manifold disasters on humanity or to try to understand geographical communalities and differences in their own right. In other words, it is past time to choose sides. But first we need to understand and overcome our own bad habits of thinking and doing: *vincit qui se vincit*.

Recommended reading

Agnew, J. A. 1993 Representing space: space, scale and culture in social science, in J. Duncan and D. Ley (eds) *Place/Culture/Representation*. London: Routledge.

Agnew, J. A. 2002 Political power and geographical scale, in Y. H. Ferguson and R. J. B. Jones (eds) *Political Space: Frontiers of Change and Governance in a Globalizing World*. Albany, NY: SUNY Press.

Anderson, J. (ed.) 2002 *Transnational Democracy: Political Spaces and Border Crossings*. London: Routledge.

Brzezinski, Z. 1997 *The Grand Chessboard: American Primacy and Its Geostrategic Imperatives*. New York: Basic Books.

Derluguian, G. M. and Greer, S. L. (eds) 2000 *Questioning Geopolitics: Political Projects in a Changing World-System*. Westport, CT: Greenwood Press.

Harvey, D. 1989 *The Condition of Postmodernity*. Oxford: Blackwell.

Morley, D. 2000 *Home Territories: Media, Mobility and Identity*. London: Routledge.

Ó Tuathail, G. 1996 *Critical Geopolitics: The Politics of Writing Global Space*. Minneapolis: University of Minnesota Press.

Glossary

cosmography   The science of the world in which the structure of the cosmos in general and the world in particular was to be accounted for in an exhaustive inventory of particular features, including folklore as well as documented observation, but always in terms of a pre-existing imagined world map that was to be filled in. Popular in the fifteenth and sixteenth centuries, it was later eclipsed by the rise of modern geography and anthropology.

discourse   The setting or domain in which words are used and take on specific meanings. In geopolitical discourse, words such as 'backward', 'race', 'heartland', and 'Great Power' take on meaning in relation to other words within a stream of understandings that provides guidance to and interpretation of practices or action.

élites   The communities of government officials, political leaders, foreign policy experts and their immediate supporters throughout the world who conduct, influence and comment upon world politics. In relation to foreign policy, such communities include both the practitioners and the 'intellectuals of statecraft'.

geopolitical imagination, modern   The view of the world and its geographical workings that accompanied the rise of the state and capitalism in Europe and that was both stimulated by and informed the European encounter with the rest of the world.

geopolitics   The study of the impact of geographical distributions and divisions on the conduct of world politics. In its original usage it referred to the impact on inter-state relations of the spatial disposition of continents and oceans and the distribution of natural and human resources. Today, however, the term also covers examination of all of the geographical assumptions, designations and understandings that enter into the making of world politics (as in critical geopolitics) and how these change in concert with material conditions (historical geopolitics).

globalization   The processes and ideas associated with a world in which information, trade, investment, finance and cultural phenomena are increasingly worldwide in incidence and less subject to effective regulation by states (particularly more powerful ones) than was formerly the case. In an extreme form this trend is seen as leading to a postmodern world of 'flows' replacing the world of 'territories' that has characterized the modern world. Even flows, however, require anchors in place, though these may not necessarily be at the geographical scale of existing states. But the term is also used to cover the emergence of supranational modes of political regulation of the world economy (the European Union, the UN, the International Monetary Fund, etc.). Terms such as 'regionalism' and 'globalism' might be more usefully applied to these developments.

Great Powers   The dominant or leading states and empires of a particular epoch.

hegemony   The dominant understandings and rules governing political and economic practice of a given period as accepted by coalitions of élites. Sometimes associated in world politics with the dominance exerted by a single state (or hegemon). This usage also accepts that dominance is rarely simply coercive but extends into creating consensus for certain policies and to control over the 'rules of the game' or agenda setting for world politics in general.

ideology   A structured understanding of how the world works and in whose interests. In one usage 'ideology' is reserved for one's political opponents, implying that it is false or misleading thought. Another sees any pattern of thought directed towards 'unmasking' the world as it appears as an ideology.

marchland   A borderland area frequently contested by adjoining states. Also can refer to a state caught between more powerful neighbouring states.

mercantilism   An economic doctrine prevalent in Europe in the sixteenth to eighteenth centuries in which states strongly regulated industry and trade on the basis of the following four beliefs: that exports to foreign countries increase national strength; that exports are preferable to internal trade and foreign imports; that state wealth depends primarily upon the possession of gold and silver; and that state intervention is justified to achieve these objectives. Mercantilist ideas and practices never entirely disappeared. They crop up in connection with most territorial modes of economic organization, including the exploitation of colonies to provide raw materials and precious metals.

nation   A group of people of common nationality (kinship, ethnicity or civic bonds being the most common basis to the definition of nationality) occupying and claiming a historic territory.

nation-state   A state in which the territorial boundaries also enclose a claimed nation which the state represents.

polity   A structured organization of power within a social group that has a distinctive identity, can mobilize group members and persists over time. By this definition not only territorially bounded states but churches, interest groups and social movements can be types of polity. Polities need not be organized territorially and can coexist in layered, overlapping and interacting webs of power and influence.

power   The capacity to achieve goals. Often, the power to coerce others (or power over) is privileged definitionally at the expense of the power of agency (or power to).

spatiality   The way in which terrestrial space is thought of as affecting the organization of a given phenomenon such as power or social relations.

state   The government of a bounded territory which claims common political and legal authority and a monopoly of legitimate force and other sovereign powers throughout its jurisdiction. This is often fused with the idea of nation (a common people) to create the nation-state, in which there is a territorial matching of nation and state.

state territoriality   The association of the state with a discrete and well-defined territory.

Three Worlds   The Three Worlds of development as defined in the Cold War: the First World of modern, capitalist states, the Second World of modern but communist states, and the Third World in which the first two vied for influence.

transnational liberalism   A set of ideas and practices associated with economic organization that is not confined by state boundaries but incorporates the whole world as a potential space for investment, trade and production.

Westphalian system   The set of mutually recognized states that slowly emerged as a result of the Treaty of Westphalia (1648) in which the right of a ruler to an exclusive territorial jurisdiction was first legislated. As new states (including old empires and other polities) have appeared over the years they have wanted to be recognized as parts of this system, even though the European original relied on a close approximation between nation and state. The increased permeability of state boundaries as the world economy globalizes, the lack of effectiveness and legitimacy of many states, and the lack of matching between nations and states all undermine the primary requirement of exclusive jurisdiction.

Bibliography

Achebe, C. 1975 *Morning Yet on Creation Day*. London: Faber.

Agnew, J. A. 1993 Representing space: space, scale and culture in social science, in J. Duncan and D. Ley (eds) *Place/Culture/Representation*. London: Routledge.

Agnew, J. A. 1993 The United States and American hegemony, in P. J. Taylor (ed.) *The Political Geography of the Twentieth Century*. London: Belhaven Press.

Agnew, J. A. 1994 The territorial trap: the geographical assumptions of international relations theory. *Review of International Political Economy*, 1: 53–80.

Agnew, J. A. 2001 *Reinventing Geopolitics: Geographies of Modern Statehood*. Heidelberg, Germany: Hettner Lectures, Institute of Geography, University of Heidelberg.

Agnew, J. A. 2001 Disputing the nature of the international in political geography. *Geographische Zeitschrift*, 89: 1–16.

Agnew, J. A. 2002 Political power and geographical scale, in Y. H. Ferguson and R. J. B. Jones (eds) *Political Space: Frontiers of Change and Governance in a Globalizing World*. Albany, NY: SUNY Press.

Agnew, J. A. and Corbridge, S. 1995 *Mastering Space: Hegemony, Territory and International Political Economy*. London: Routledge.

Alker, H. 1996 *Rediscoveries and Reformulations: Humanistic Methodologies for International Studies*. Cambridge: Cambridge University Press.

Allen, J. 2002 *Lost Geographies of Power*. Oxford: Blackwell.

Anderson, J. (ed.) 2002 *Transnational Democracy: Political Spaces and Border Crossings*. London: Routledge.

Arrighi, G. and Silver, B. (eds) 1999 *Chaos and Governance in the Modern World-System*. Minneapolis: University of Minnesota Press.

Avishai, B. 1995 The inseparables: for Israel, separation from the Palestinians is a chimera. *New Yorker*, 16 October: 5–6.

Banfield, E. 1958 *The Moral Basis of a Backward Society*. New York: Free Press.

Barnouw, D. 1990 *Visible Spaces: Hannah Arendt and the German-Jewish Experience*. Baltimore, MD: Johns Hopkins University Press.

Bassin, M. 1987 Race contra space: the conflict between German *Geopolitik* and National Socialism. *Political Geography Quarterly*, 6: 115–34.

Bassin, M. 1991 Russia between Europe and Asia: the ideological construction of geographical space. *Slavic Review*, 50: 1–17.

Bassin, M. 1993 Turner, Solov'ev and the 'frontier hypothesis': the nationalist signification of open spaces. *Journal of Modern History*, 65: 473–511.

Berger, J. 1972 *Ways of Seeing*. London: BBC Books.

Biersteker, T. 1993 Evolving perspectives on international political economy: twentieth-century contexts and discontinuities. *International Political Science Review*, 14: 7–33.

Biersteker, T. and Weber, C. (eds) 1996 *State Sovereignty as Social Construct*. Cambridge: Cambridge University Press.

Brantlinger, P. 1996 *Fictions of State: Culture and Credit in Britain, 1694–1994*. Ithaca, NY: Cornell University Press.

Brotton, J. 1998 *Trading Territories: Mapping the Early Modern World*. Ithaca, NY: Cornell University Press.

Brzezinski, Z. 1986 *Game Plan*. Boston: Atlantic Books.

Brzezinski, Z. 1997 *The Grand Chessboard: American Primacy and Its Geostrategic Imperatives*. New York: Basic Books.

Burleigh, M. 1988 *Germany Turns Eastward: A Study of* Ostforschung *in the Third Reich*. Cambridge: Cambridge University Press.

Campbell, D. 1992 *Writing Security: United States Foreign Policy and the Politics of Identity*. Minneapolis: University of Minnesota Press.

Campbell, D. 1997 *National Deconstruction: Violence, Identity and Justice in Bosnia*. Minneapolis: University of Minnesota Press.

Caraley, D. J. (ed.) 2002 *September 11, Terrorist Attacks, and U.S. Foreign Policy*. New York: Academy of Political Science.

Chace, J. 2002 Imperial America and the Common Interest. *World Policy Journal*, 19: 1–9.

Chow, R. 1996 We endure, therefore we are: survival, governance, and Zhang Yimou's *To Live*. *South Atlantic Quarterly*, 95: 1039–64.

Cohen, B. J. 1977 *Organizing the World's Money*. New York: Basic Books.

Cohen, B. J. 1998 *The Geography of Money*. Ithaca, NY: Cornell University Press.

Collins, R. 1986 *Weberian Sociological Theory*. Cambridge: Cambridge University Press.

Connolly, W. 1989 *Political Theory and Modernity*. Oxford: Blackwell.

Constantine, S. 1986 *Buy and Build: The Advertising Posters of the Empire Marketing Board*. London: HMSO.

Cosgrove, D. 2001 *Apollo's Eye: A Cartographic Genealogy of the Earth in the Western Imagination*. Baltimore: Johns Hopkins University Press.

Cowhey, P. F. and Aronson, J. D. 1993 *Managing the World Economy: The Consequences of Corporate Alliances*. New York: Council on Foreign Relations Press.

Cox, M. 1990 From the Truman doctrine to the second superpower détente: the rise and fall of the Cold War. *Journal of Peace Research*, 27: 25–41.

Cox, R. W. 1987 *Power, Production, and World Order: Social Forces in the Making of History*. New York: Columbia University Press.

Cox, R. W. 1992 Multilateralism and world order. *Review of International Studies*, 18: 161–80.

Crocker, C. 1977 The social function of rhetorical forms, in J. D. Sapir and C. Crocker (eds) *The Social Uses of Metaphor: Essays on the Anthropology of Rhetoric*. Philadelphia: University of Pennsylvania Press.

Crone, G. R. 1978 *Maps and Their Makers: An Introduction to the History of Cartography*. Hamden, CT: Archon Books.

Croxton, D. 1999 The Peace of Westphalia of 1648 and the origins of sovereignty. *International History Review*, 21: 569–91.

Cumings, B. 1999 Still the American Century. *Review of International Studies*, 25: 271–99.

Curtis, L. 1983 *Nothing but the Same Old Story*. London: Information on Ireland.

Darwin, C. 1839 *Journal of Researches into the Geology and Natural History of the Various Countries Visited by the H.M.S. Beagle*. London: Henry Colburn.

Davidson, B. 1992 *The Black Man's Burden: Africa and the Curse of the Nation-State*. New York: Times Books.

Department of Defense (DoD) *1995 Annual Report*. Washington, DC: US Government Printing Office.

Derluguian, G. M. and Greer, S. L. (eds) 2000 *Questioning Geopolitics: Political Projects in a Changing World-System*. Westport, CT: Greenwood Press.

Desai, M. 2002 *Marx's Revenge: The Resurgence of Capitalism and the Death of Statist Socialism*. London: Verso.

Deudney, D. 1995 Nuclear weapons and the waning of the real-state. *Daedalus*, 124: 209–31.

Deudney, D. 1996 Binding sovereigns: authorities, structures, and the geopolitics of the Philadelphian system, in T. Biersteker and C. Weber (eds) *State Sovereignty as Social Construct*. Cambridge: Cambridge University Press.

Deudney, D. 2000 Regrounding realism: anarchy, security, and changing material contexts. *Security Studies*, 10: 1–45.

Dijkink, G. 1996 *National Identity and Geopolitical Visions: Maps of Pride and Pain*. London: Routledge.

Durand, M. F., Lévy, J. and Retaillé, D. 1992 *Le Monde: espaces et systèmes*. Paris: Presses de la Fondation Nationale des Sciences Politiques.

Eagleton, T. 1991 *Ideology: An Introduction*. London: Verso.

Eco, U. 1999 *Serendipities: Language and Lunacy*. New York: Columbia University Press.

Economist, The 2002 Muslim opinion polls. 19 October: 43.

Elliott, J. H. 1991 The world after Columbus. *New York Review of Books*, 10 October: 10–14.

Fabian, J. 1983 *Time and Its Other: How Anthropology Makes Its Object*. New York: Columbia University Press.

Ferguson, Y. H. and Mansbach, R. W. 1996 *Polities: Authority, Identities, and Change*. Columbia: University of South Carolina Press.

Frank, R. H. and Cook, P. J. 1995 *The Winner-Take-All Society*. New York: Free Press.

Friedberg, A. L. 1988 *The Weary Titan: Britain and the Experience of Relative Decline, 1895–1905*. Princeton, NJ: Princeton University Press.

Friedrich, P. 1989 Language, ideology, and political economy. *American Anthropologist*, 91: 295–312.

Gervasi, T. 1988 *Soviet Military Power: The Pentagon's Propaganda Document Annotated and Corrected*. New York: Vintage.

Gilman, S. L. 1992 Plague in Germany, 1939/1989: cultural images of race, space, and disease, in A. Parker, M. Russo, D. Summer and P. Yaeger (eds) *Nationalism and Sexualities*. London: Routledge.

Gilpin, R. 1981 *War and Change in World Politics*. Cambridge: Cambridge University Press.

Ginsborg, P. 1990 *A History of Contemporary Italy: Society and Politics, 1943–1988*. London: Penguin.

Goody, J. 1996 *The East in the West*. Cambridge: Cambridge University Press.

Gopnik, A. 1996 The first Frenchman. *New Yorker*, 7 October: 44–53.

Gorbachev, M. S. 1987–88 The reality and guarantees of a secure world. Foreign Broadcast Information Service SOV – 87–180, 17 September 1987.

Gowan, P. 1999 *The Global Gamble: Washington's Faustian Bid for Global Dominance*. London: Verso.

Gowan, P. 2002 A calculus of power. *New Left Review*, 16: 47–67.

Grafton, A. 1992 *New Worlds, Ancient Texts: The Power of Tradition and the Shock of Discovery*. Cambridge, MA: Belknap Press of the Harvard University Press.

Gramsci, A. 1971 *Selections from the Prison Notebooks*. London: Lawrence and Wishart.

Gregory, D. 1994 *Geographical Imaginations*. Oxford: Blackwell.

Gregory, R. 1978 The domino theory, in A. DeConde (ed.) *Encyclopedia of American Foreign Policy*, vol. 1. New York: Charles Scribner's Sons.

Gress, D. 1998 *From Plato to NATO: The Idea of the West and Its Opponents*. New York: Free Press.

Grunberg, I. 1990 Exploring the 'myth' of hegemonic stability. *International Organization*, 44: 431–77.

Guha, R. 1989 Dominance without hegemony and its historiography. *Subaltern Studies*, 6: 210–309.

Hardt, M. and Negri, A. 2000 *Empire*. Cambridge, MA: Harvard University Press.

Harley, J. B. 1989 Deconstructing the map. *Cartographica*, 26: 1–20.

Harvey, D. 1989 *The Condition of Postmodernity*. Oxford: Blackwell.

Haslam, J. 2002 *No Virtue Like Necessity: Realist Thought in International Relations since Machiavelli*. New Haven, CT: Yale University Press.

Hearn, C. G. 2002 *Tracks in the Sea: Matthew Fontaine Maury and the Mapping of the Oceans*. Camden, ME: International Marine/McGraw-Hill.

Heffernan, M. 1998 *The Meaning of Europe: Geography and Geopolitics*. London: Arnold.

Hegel, G. W. F. 1821/1967 *Philosophy of Right*. Oxford: Oxford University Press.

Hegel, G. W. F. 1931 *The Phenomenology of Mind*, 2nd English edition. New York: Humanities Press.

Heidegger, M. 1959 *An Introduction to Metaphysics*. New Haven, CT: Yale University Press.

Helleiner, E. 1994 *States and the Reemergence of Global Finance: From Bretton Woods to the 1990s*. Ithaca, NY: Cornell University Press.

Helleiner, E. 1999 Historicizing territorial currencies: monetary space and the nation-state in North America. *Political Geography*, 18: 309–39.

Helms, M. 1988 *Ulysses Sail: An Ethnographic Odyssey of Power, Knowledge, and Geographical Distance*. Princeton, NJ: Princeton University Press.

Higonnet, M. R. 1994 New cartographies, an introduction, in M. R. Higonnet and J. Templeton (eds) *Reconfigured Spheres: Feminist Explorations of Literary Space*. Amherst: University of Massachusetts Press.

Holzman, F. D. 1989 Politics and guesswork: the CIA and DIA estimates of Soviet military spending. *International Security*, 13: 101–31.

Huntington, S. P. 1993 The clash of civilizations? *Foreign Affairs*, 72: 22–49.

Ikenberry, G. J. 2002 America's imperial ambitions. *Foreign Affairs*, 81: 44–60.

Inayatullah, N. 1997 Theories of spontaneous disorder. *Review of International Political Economy*, 4: 319–48.

Jacobson, D. 1996 *Rights across Borders: Immigration and the Decline of Citizenship*. Baltimore: Johns Hopkins University Press.

Jay, M. 1993 *Downcast Eyes: The Denigration of Vision in Twentieth Century French Thought*. Berkeley and Los Angeles: University of California Press.

Julius, D. 1990 *Global Companies and Public Policy: The Growing Challenge of Foreign Direct Investment*. New York: Council on Foreign Relations.

Kahler, M. 1992 Multilateralism with small and large numbers. *International Organization*, 46: 681–708.

Kearns, G. 1993 Prologue: fin de siècle geopolitics: Mackinder, Hobson and theories of global closure, in P. J. Taylor (ed.) *The Political Geography of the Twentieth Century*. London: Belhaven Press.

Kennan, G. [Mr X] 1947 The sources of Soviet conduct. *Foreign Affairs*, 25: 566–82.

Kennedy, P. 1987 *The Rise and Fall of the Great Powers: Economic Change and Military Conflict from 1500 to 2000*. New York: Random House.

Kennedy, P. 1995 'Too serious a business', a reply to Professor Krugman. *Peace Economics, Peace Science and Public Policy*, 2: 16–22.

Keohane, R. O. 1984 *After Hegemony: Cooperation and Discord in the World Political Economy*. Princeton, NJ: Princeton University Press.

Kermode, F. 1967 *The Sense of an Ending*. New York: Oxford University Press.

Kern, S. 1983 *The Culture of Time and Space, 1880–1918*. Cambridge, MA: Harvard University Press.

Klare, M. 2002 Hawks with their eyes on oil. United States: energy and strategy. *Le Monde Diplomatique* (English-language edition), November: 1–2.

Kristof, L. 1968 The Russian image of Russia, in C. A. Fisher (ed.) *Essays in Political Geography*. London: Methuen.

Krugman, P. 1995 A reply. *Peace Economics, Peace Science and Public Policy*, 2: 26–30.

Krugman, P. 1996 *Pop Internationalism*. Cambridge, MA: MIT Press.

Kuhl, S. 1994 *The Nazi Connection: Eugenics, American Racism, and German National Socialism*. New York: Oxford University Press.

Lanaro, S. 1989 *L'Italia nuova. Identità e sviluppo, 1861–1988*. Turin: Einaudi.

Lestringant, F. 1994 *Mapping the Renaissance World: The Geographical Imagination in the Age of Discovery*. Berkeley and Los Angeles: University of California Press.

Lewis, M. and Wigen, K. 1997 *The Myth of Continents*. Berkeley and Los Angeles: University of California Press.

Lieven, A. 2002 The push for war. *London Review of Books*, 3 October: 8–11.

Lugard, F. D. 1926 *The Dual Mandate in Tropical Africa*. Edinburgh: Oliver and Boyd.

Luke, T.W. 2003 Postmodern geopolitics: the case of the 9.11 terrorist attacks, in J. Agnew, K. Mitchell and G. Toal (eds) *A Companion to Political Geography*. Oxford: Blackwell.

Mackinder, H. J. 1904 The geographical pivot of history. *Geographical Journal*, 13: 421–37.

McNamara, R. J. 1995 *In Retrospect: The Tragedy and Lessons of Vietnam*. New York: Times Books.

McNeill, W. H. 1982 *The Pursuit of Power*. Chicago: University of Chicago Press.

Mahan, A.T. 1965 *The Influence of Sea Power upon History, 1660–1783*. London: Methuen (original edition Boston, 1890).

Manchester, W. 1992 *A World Lit Only by Fire: The Medieval Mind and the Renaissance, Portrait of an Age*. Boston: Little, Brown.

Mandrou, R..1978 *From Humanism to Science, 1480–1700*. Harmondsworth, UK: Penguin.

Mann, M. 1984 The autonomous power of the state: its origins, mechanisms and results. *European Journal of Sociology*, 25: 185–213.

Mason, T. 1988 Italy and modernization: a montage. *History Workshop*, 25/26: 127–47.

Mearsheimer, J. J. 2001 *The Tragedy of Great Power Politics*. New York: Norton.

Mertes, T. 2002 Grass-roots globalism. *New Left Review*, 17: 101–10.

Mitchell, T. L. 1839 *Three Expeditions into the Interior of Eastern Australia: with descriptions of the Recently Explored Regions of Australia Felix and of the Present Colony of New South Wales*, 2 vols, 2nd edition. London: T. W. Boone.

Money, J. 1997 No vacancy: the political geography of immigration control in advanced industrial countries. *International Organization*, 51: 685–720.

Morley, D. 2000 *Home Territories: Media, Mobility and Identity*. London: Routledge.

Ninkovich, F. 1994 *Modernity and Power: A History of the Domino Theory in the Twentieth Century*. Chicago: University of Chicago Press.

Nordhaus, W. D. 2002 Iraq: the economic consequences of war. *New York Review of Books*, 49, 5 December: 9–12.

Nussbaum, F. 1995 *Torrid Zones: Maternity, Sexuality, and Empire in Eighteenth-Century English Narratives*. Baltimore, MD: Johns Hopkins University Press.

Nye, J. 1990 The changing nature of world power. *Political Science Quarterly*, 105: 177–92.

Ó Tuathail, G. 1993 Japan as threat: geo-economic discourses on the US–Japan relationship in US civil society, 1987–1991, in C. H. Williams (ed.) *The Political Geography of the New World Order*. London: Belhaven Press.

Ó Tuathail, G. 1996 *Critical Geopolitics: The Politics of Writing Global Space*. Minneapolis: University of Minnesota Press.

Ó Tuathail, G. and Agnew, J. A. 1992 Geopolitics and discourse: practical geopolitical reasoning in American foreign policy. *Political Geography*, 11: 190–204.

O'Loughlin, J. and Grant, R. 1990 The political geography of presidential speeches, 1946–1987. *Annals of the Association of American Geographers*, 80: 504–30.

Pagden, A. 1992 *European Encounters with the New World: From Renaissance to Romanticism*. New Haven: Yale University Press.

Pagden, A. 1994 *Lords of All the World: Ideologies of Empire in Spain, Britain and France c. 1500–c. 1800*. New Haven: Yale University Press.

Pagden, A. 2001 *Peoples and Empires: Europeans and the Rest of the World from Antiquity to the Present*. London: Weidenfeld and Nicolson.

Perkins, D. 1992 *Is Literary History Possible?* Baltimore: Johns Hopkins University Press.

Pletsch, C. E. 1981 The Three Worlds, or the division of social scientific labor, circa 1950–1975. *Comparative Studies in Society and History*, 23: 565–90.

Polanyi, K. 1944 *The Great Transformation*. Boston: Beacon Press.

Porter, M. 1990 *The Competitive Advantage of Nations*. London: Macmillan.

Pratt, J. W. 1935 The ideology of American expansion, in A. Craven (ed.) *Essays in Memory of William E. Dodd*. Chicago: University of Chicago Press.

Record, J. 2002 *Making War, Thinking History: Munich, Vietnam, and Presidential Uses of Force from Korea to Kosovo*. Washington, DC: Naval Institute Press.

Reich, R. 1991 The myth of 'Made in the U.S.A.'. *Wall Street Journal*, 5 July: A6.

Reich, R. 1996 *The Work of Nations: Preparing Ourselves for 21st Century Capitalism*. New York: Knopf.

Renan, E. 1992 [1882] Qu'est-ce qu'une nation?, in J. Roman (ed.) *Discours et conférences, Œuvres complètes*, vol. 1. Paris: Pocket.

Roberts, D. 2001 *Great Exploration Hoaxes*. New York: Modern Library.

Rosenberg, E. S. 1982 *Spreading the American Dream: American Economic and Cultural Expansion, 1890–1945*. New York: Hill and Wang.

Rostow, W. 1960 *The Stages of Economic Growth: A Non-communist Manifesto*. New York: Cambridge University Press.

Ruggie, J. G. 1993 Territoriality and beyond: problematizing modernity in international relations. *International Organization*, 47: 139–74.

Ryan, M. T. 1981 Assimilating new worlds in the sixteenth and seventeenth centuries. *Comparative Studies in Society and History*, 23: 519–38.

Ryan, S. 1996 *The Cartographic Eye: How Explorers Saw Australia*. Cambridge: Cambridge University Press.

Sader, E. 2002 Beyond civil society: the Left after Porto Alegre. *New Left Review*, 17: 87–99.

Said, E. W. 1978 *Orientalism: Western Conceptions of the Orient*. New York: Vintage.

Sandholtz, W. 1993 Choosing union: monetary politics and Maastricht. *International Organization*, 47: 1–39.

Schmitter, P. 1971 Still the century of corporatism? *Review of Politics*, 36: 85–131.

Schwoebel, R. 1967 *The Shadow of the Crescent: The Renaissance Image of the Turk (1453–1517)*. Nieukoop: De Graaf.

Scott, A. J. 1996 Regional motors of the global economy. *Futures*, 28: 391–411.

Sewell, W. H. 1996 Three temporalities: toward an eventful sociology, in T. J. McDonald (ed.) *The Historic Turn in the Human Sciences*. Ann Arbor: University of Michigan Press.

Shapiro, M. J. 1989 Representing world politics: the sport/war intertext, in J. Der Derian and M. J. Shapiro (eds) *International/Intertextual Relations*. Lexington, MA: Lexington Books.

Shapiro, M. J. 1995 The ethics of encounter: unreading/unmapping the Imperium. Paper presented at the International Studies Association Annual Meeting, Chicago, February.

Shapiro, M. J. 1999 Samuel Huntington's moral geography. *Theory and Event*, 2/4: 1–11.

Sharp, J. P. 2000 *Condensing Communism: 'Reader's Digest' and American Identity*. Minneapolis: University of Minnesota Press.

Signorelli, A. 1986 Review of *La nostra Italia*. *L'Indice*, 8: 46.

Simpson, D. 1996 *The Academic Postmodern and the Rule of Literature: A Report on Half-Knowledge*. Chicago: University of Chicago Press.

Smith, A. D. 1991 *National Identity*. London: Penguin.

Smith, G. 1994 *The Last Years of the Monroe Doctrine, 1945–1993*. New York: Hill and Wang.

Soros, G. 1998–99 Capitalism's last chance? *Foreign Policy*, 113: 66.

Spanos, W. V. 2000 *America's Shadow: An Anatomy of Empire*. Minneapolis: University of Minnesota Press.

Spengler, O. 1926 *The Decline of the West*. New York: Knopf.

Springborg, P. 1992 *Western Republicanism and the Oriental Prince*. Cambridge: Polity Press.

Spruyt, H. 1994 *The Sovereign State and Its Competitors: An Analysis of Systems Change*. Princeton, NJ: Princeton University Press.

Spurr, D. 1993 *The Rhetoric of Empire: Colonial Discourse in Journalism, Travel Writing, and Imperial Administration*. Durham, NC: Duke University Press.

Stephanson, A. 1995 *Manifest Destiny: American Expansion and the Empire of Right*. New York: Hill and Wang.

Stokes, E. 1959 *The English Utilitarians and India*. London: Oxford University Press.

Taylor, P. J. (ed.) 1993 *The Political Geography of the Twentieth Century*. London: Belhaven Press.

Teschke, B. 2002 Theorizing the Westphalian system of states: international relations from absolutism to capitalism. *European Journal of International Relations*, 8: 5–48.

Tiley, R. 2001 *Australian Navigators: Picking Up Shells and Catching Butterflies in an Age of Revolution*. Sydney: Kangaroo Press.

Tolan, S. 2002 Beyond regime change. The administration doesn't simply want to oust Saddam Hussein. It wants to redraw the map of the Middle East. *Los Angeles Times*, 1 December: M1–6.

Toulmin, S. 1990 *Cosmopolis: The Hidden Agenda of Modernity*. Chicago: University of Chicago Press.

Tullio-Altan, C. 1986 *La nostra Italia. Arretratezza socioculturale, clientelismo, trasformismo e ribellismo dall'Unità ad oggi*. Milan: Feltrinelli.

Wade, R. 1998–99 The fight over capital flows. *Foreign Policy*, 113: 41–54.

Walker, R. B. J. 1993 *Inside/Outside: International Relations as Political Theory*. Cambridge: Cambridge University Press.

Wallerstein, I. 1991 *Unthinking Social Science: The Limits of Nineteenth-Century Paradigms*. Cambridge: Polity Press.

Waltz, K. E. 1979 *Theory of International Politics*. New York: Random House.

Weber, M. 1978 *Economy and Society*, vol. 1, eds G. Roth and C. Wittich. Berkeley and Los Angeles: University of California Press.

Whitfield, P. 1998 *New Found Lands: Maps in the History of Exploration*. New York: Routledge.

Wills, G. 1992 *Lincoln at Gettysburg: The Words That Remade America*. New York: Simon and Schuster.

Wolf, E. R. 1982 *Europe and the People without History*. Berkeley and Los Angeles: University of California Press.

Wolin, S. 1985 Postmodern society and the absence of myth. *Social Research*, 52: 217–39.

Zerubavel, E. 1992 *Terra Cognita: The Mental Discovery of America*. New Brunswick, NJ: Rutgers University Press.

Index

Note: Page numbers in italic refer to illustrations.

Aborigines 37, 38, 39–40
'absolute individuation' 59
Achebe, Chinua 90
Acton, Lord 96
'advanced' countries 35, 36; *see also*
    modernity
Afghanistan 1, 2, 108, 121
Africa 23, 26, 27, 56–7; and colonialism 9,
    45, 99; maps of 18, 25; South Africa 80
ages of geopolitics 12, 75, 85–113
Agnew, J.A. 106, 110
al-Qaeda 1, 4, 120–22
America *see* American colonies; Central
    America; United States
American colonies: maps of 19
'amoral familism' 41–2
anarchy: international 69–70, 72–4
Angola 111
anti-geopolitics 116
anti-Semitism 95–7, 101
appeasement: US fear of 8, 107, 111, 112
Aristotle 53
Aronson, J.D. 59, 60
Asia 23, 24–5, 26, 27–8, 44–5, 89
Augustine, St 69
Australia: European encounter 11, 35,
    37–40; exploration of 11, 47, 48; maps of
    19, 38; politics in 38; primitiveness of 47,
    48
Avishai, B. 81
'axis of evil' *4*

'backwardness' 8, 11, 35–48, 89, 101, 108;
    of Italy 40–2; and Third World 108
Bali 2
Banfield, Edward 41
Barnouw, D. 96

Bassin, M. 88, 90
Berger, John 21
Berlin: divided 2, *107*
Biersteker, T. 101
binary geographies 11, 23–31
bin Laden, Osama 1, 4, 120–22
bipolarity of superpowers 102, 109–12
Blaeu (Coesius, Willem Janszoon) 19
Bolshevik Revolution (1917) 104
Bosnia 7, 93
boundaries of states, natural 98–9
Brantlinger, P. 63
Braveheart (film) 88
Bretton Woods Agreement (1944)/system
    76
Britain: colonial economic benefits 87, 89,
    93, 94; colonialism 28, 45; dominant
    economy 87, 93; as hegemonic state 22,
    57, 75–6, 87, 93, 112 ; racial superiority
    95–6; social unrest 98; *see also* England;
    Ireland; Scotland
Britannia 27
Brunelleschi, Filippo 21
Brzezinski, Zbigniew 71, 109, 128
'Bush doctrine' 4–5
Bush, George H.W. 71
Bush, George W. 1, 4, 5, 123, 125

Canadian Indians *91*
capitalism 26–8, 101; global/world 59, 87,
    118; industrial 75, 86; informational 12,
    31; merchant 59; United States epitomizes
    105
cartography *see* maps
Celtic races 96
Central America 93, 110, 111
Central Europe 2–3, 4

Ceylon 93
change: political 12
Child, G. 27
Chile 111
China 30, 46, 103, 104, 118, 123–4
Chorography 20
Christendom 23–4, 44, 88
Christianity 23–4, 43; Satanic
   representations 70; and war 69
Churchill, Sir Winston 107
city-states 11, 42
'civic maturity' 41–2
civilizational geopolitics 12, 23, 85, 86–93
classic empires 11
classical system of state organization 73
clientelismo 41
climatic zones 26–7, 32
Clinton administration 123–4
'closed world' concept 99
Clovis, king of the Franks 26
coercive power 55–8
Cohen, B. 63
Cold War 1, 2, 3, 4, 8, 10, 12, 29, 47, 57,
   76, 115, 119, 122, 124, 128, 131; and
   binary divisions 30–1; ideological
   geopolitics 12, 102–12; nostalgia for 122;
   and world economy 61
colonialism: in Africa 9, 45, 56–7, 99; and
   Australia 37–40; and competition for
   primacy 76; and currency 64;
   decolonization 9, 56–7, 105; economics
   of 6, 11, 28, 30, 87, 88, 90, 94, 95,
   99; European 35, 88, 89–92; maps
   reinforce 19; of superpowers 109; view
   of natives 89, 91, 95–6; *see also*
   imperialism
Columbus, Christopher 18, 42, 48
communism 1, 100, 103, 106, 111; Soviet
   Union personifies 106; threat of 1, 8
competition: inter-state 12, 67–83
Concert of Europe 56, 87, 92, 93, 99
confederations 11
Congress of Berlin (1884–5) 99
containment doctrines 7–8, 29–30, 106,
   110–11
Contarini, Giovanni 18
'continental' states 101
continuity in geopolitical imagination 85
Contras (Nicaragua) 106, 111
cooperation benefits 54, 78, 79, 81, 112
Cosgrove, D.E. 20
cosmography 17–20, 31

Council for Mutual Economic Assistance
   (CMEA) 109
Cowhey, P.F. 59, 60
Cox, M. 104, 107
Cox, R.W. 82
credibility 8, 111
Croatia 7
Crocker, C. 28
cross-border flows 57–9, 60, 78–9, 81–2,
   116–18, 120, 124–5
Cuba 111
culture: Aboriginal 39; culture wars 119–22;
   world culture 82
currency 63–4; and colonialism 64;
   deterritorialization of 61; dollar
   domination 103; local 64
Czechoslovakia 2–3

Dante Alighieri 30
Darwin, Charles 95
decolonization 9, 30, 101
de Gaulle, Charles 58
Depression (1930s) 76, 100
despotic power 55
Deudney, D. 58, 72
Devil *see* Satan
Diaz, Bartolomeu 18
disciplinary geopolitics 128
division of labor 26, 75
'domestic' affairs 51, 53, 54
domino effect 8, 29–30, 103, 111

Eastern Europe 25, 103, 108
East/Orient 24–5, 27, 32, 39, 44–5
East–West dichotomy *see* binary
   geographies
economic liberalism 100, 103
economic nationalism 98, 99–100
economics: of colonialism 16, 28, 30, 87,
   88, 90, 94, 95, 99; currency 63–4;
   economic growth 78, 86–7, 99, 103;
   'market access' economy 59–60, 61,
   78–9, 82, 116–17, 118; *see also* world
   economy
educating state 52
Eisenhower, Dwight D. 111
Elliot, H.J. 17
English nationalism 92
Enlightenment 22, 45
environmental determinism 101
essentialization of places 36

ethnic conflicts/separatism 7, 80–1
ethnocentricity of maps 19
eugenics 96
eurodollar markets 64
Europe: and binary divisions 23–31; and
    civilizational geopolitics 12, 86–93;
    colonialism 35, 88, 89–91, 131;
    competition in state system 73; East–West
    opposition 44–5; Eurocentric view 9, 19,
    35, 88 idealized past 36; identified with
    Christendom 23–4; maps of 23, 88, 92;
    medieval political organization 72; model
    for modernity 11, 42–4, 46; nationalism in
    51–2; period of primacy 75; postwar 107,
    111; territorial states 11–12; vision of
    world 15, 16, 17, 32; voyages of
    discovery 16–18, 19; see also Concert of
    Europe; Eastern Europe; Renaissance
European Union 1–3, 9, 52, 62, 63, 72, 78,
    120; monetary union plans 64
evolutionary theories 95–6, 97–8
exoticization 36, 39, 44–5
exploration: of Australia 11, 54, 37–40, 47,
    48; voyages of discovery 11, 19, 88

fascism: Italian 40, 100
feminist critique 15
Ferguson, Adam 27
Fichte, Johann Gottlieb 97
fiction 47–8
'fifth column' 8
financial markets 58, 64
'Finlandization' 109
First World War 93
flows see cross-border flows
'foreign' affairs 51, 53, 54
foreign currency transactions 63, 64
foreign direct investment 79
France 7, 24, 62, 64, 68, 76, 87, 93, 123
French Revolution 87
Friedberg, A.L. 100
Friedrich, Paul 102
frigid zones 26

G-8 summits 59
Galileo Galilei 16
General Agreement on Tariffs and Trade
    (GATT)/World Trade Organization
    (WTO) 76, 88
General Motors 79
geographical determinism 98, 101, 132
geographical identities 112

geographical location 28, 79, 101
geographical scales 6, 7, 13, 128
geography 20
geopolitical codes 110
geopolitical imagination 9–10, 12, 13, 112,
    127; continuity and periodization in
    85–6; future of 13, 115–26, 127–32
geopolitical models 10, 29–30, 98, 101,
    110
geopolitical reductionism 106
geopolitical schemas 16, 28
geopolitics: ages of geopolitics 12, 75,
    85–113, 115–26; anti-geopolitics 116;
    future of 115–26; origin and
    definition 2–7; see also civilizational
    geopolitics; ideological geopolitics;
    'linguacultural' geopolitics; naturalized
    geopolitics
Germany 7, 42, 45, 77, 87, 93, 94, 99, 100,
    101, 102; see also Nazi regime
Gervasi, T. 105
Gilpin, Robert 70
Ginsborg, Paul 40–1
global capitalism 59, 87, 116–18
global economy see world economy
global interdependence 80–1
globalization 69, 115–19, 120, 124, 131;
    and effect on primacy 76–8, 79–80
global space: and time 35–49, 99–100;
    visualized 11, 15–33
Gorbachev, Mikhail 105
Gramsci, Antonio 57
Grant, R. 110
Great Chain of Being 17, 23
Great Powers 10, 23, 28, 51, 70, 75;
    balance of power between 54, 55;
    hegemony of 9–10; in nineteenth-century
    Europe 87–8; pursuit of primacy 67–83;
    social origins of 67–9
Greece, ancient: as cultural referent 39, 44,
    88, 89–92; geographers of 88;
    organization of states in 73; philosophy
    of 68
Greek Civil War (1944–9) 106, 110, 111
Gregory, D. 22
Grey, George 39–40
Grunberg, I. 112
Guha, Ranajit 46

Haig, Alexander 109
Hanseatic League 56
Haushofer, Karl 96

'Heartland': Mackinder's model 28-*9*, 95, 98–99, 101
heaven 17, 23
Hegel, Georg Wilhelm Friedrich 45, 67, 70, 97
hegemony 9, 57–8; hegemonic stability 103, 111, 112, 113; and periodization 85; of political elites 9–10; and primacy 75; suzerainties 72; transnational liberal 61
Heidegger, Martin 97
hell 17, 30
Herder, Johann Gottfried 22
historical geography of primacy 67, 74–81, 113
Hitler, Adolf 72
Hobbes, Thomas 68, 70
Hobson, John Atkinson 99
Holy Roman Empire 56
Holzman, F.D. 105
Huntington, Samuel 119–20
Hussein, Saddam 72, 123, 124

idealization: of past 35, 92; of rivals 105
identities: geographical 112; national 92, 93, 104, 129; political 61; of states 68, 87; transnational 61–3
ideological geopolitics 12, 85, 102–12, 115
Ikenberry, G. John 124
imperial cosmologies 14
imperialism 35, 68, 88, 89, 95; and economic growth 99; rivalries in 76, 105; *see also* colonialism
India 9, 36, 68, 89
Indians, Canadian *91*
industrial capitalism 75
informational capitalism 12, 31
informational technologies 12, 58
information warfare 77
infrastructural power 55, 56
interdependence: global 80–1
international anarchy 69–70, 72–4
internationalism 104; socialist 100
internationalization: of states 78; of trade 79–80
international law 67
International Monetary Fund (IMF) 53, 76, 109, 118
international relations: coercive power in 57–8; social nature of 80; theories of 51, 54, 70
Internet as weapon 77

inter-state competition 9, 67–83
Iran 4, 77
Iraq 1, 4, 77, 123, 125
Ireland 88, 96
Iroquois Confederation 56
isolationism 104
Israel–Palestine conflict 81
Italy 4, 35, 40–2, 44, 47, 48; fascism in 40, 100; revisionist 93; unification of 40, 41; *see also* Renaissance; Roman Empire

Japan 28, 70–1, 76, 79, 102, 119, 123, 124
Jay, Martin 15
Jews 23, 96–7, 101
'just war' doctrines 69

Kant, Immanuel 45
Kearns, G. 100
Kennan, George 110
Kennedy, Paul 70, 78
Keohane, Robert 54, 57
Kermode, Frank 47
Kern, S. 99
Keynes, John Maynard 100
kin-based polities 11
Kissinger, Henry 109, 112
Kjellén, Rudolf 5, 6, 99
knowledge and perspective 21–22
Korea, North 4, 77
Korean War 111
Krugman, Paul 78

labour, division of 26, 75
Lanaro, S. 42
land powers 28
Lebensraum 99
Lestringant, F. 17
liberalism: economic 100, 103; transnational 10, 61–4, 116
Lieven, Anatol 123–24
Lincoln, Abraham: Gettysburg address 98
'linguacultural' geopolitics 102, 103, 106
linkage 8, 16, 30
local conflicts 37, 109, 111
local currencies 64
local knowledge 108–9, 127
location: geographical 28, 77, 101
Lugard, Lord 45

Machiavelli, Niccolò 24, 42, 53, 70, 74
Mackinder, Halford 6, 98–9, 125; geopolitical model 29, 95, 110

McNamara, Robert S. 8
Magellan, Ferdinand 16–17
Mahan, Alfred Thayer 68
Malaya 93
Manchester, W. 17
Mandrou, R. 44
Mann, Michael 55
maps 15, 16, 17–19, 23, 24, 120; of Europe 24, 88, 89; Ptolemy's 17–19, 36
'marchland' states 4
'maritime' states 101
'market access' economy 59–60, 61, 78–9, 80, 116–18
Martellus, Henricus 18
Maury, Matthew Fontaine 100
Mazzini, Giuseppe 41
mercantilism 76, 78
Mercator map 19, 101
merchant capitalism 59
metaphors: sporting 71–2; temporal 85
'methodological nationalism' 52, 65
militarism 98
military alliances 78
military organization of power 77
military viability of states 58
Mitchell, T.L. 39
modernity 11, 36–48, 68, 85, 103, 108, 115; in Italy 40–2; Western model for 8, 11, 42–7, 48
money see currency; economics
Monroe, James 105
Montesquieu, Charles de Secondat, Baron de la Brède et de 26
multipolarity of power 73
Munster, Sebastian 17, 89
myth 46–7

NAFTA 78
Napoleon I (Napoleon Bonaparte) 72, 92
Nasser, Gamal Abdul 108
'national development' 32, 44; in Italy 40, 42
national identities 93, 94, 116
nationalism 56, 87; economic 98, 99–100; in Europe 52–3, 92; growth of 93–5; 'methodological' 52, 65; model of past 98
national security (US) 105, 107, 111
nation-states 45, 51, 56, 87; demand sacrifice 93–4; developing identities 89, 92; economic nationalism 98, 100; rise of 52
natives: colonial view of 88, 89, 92, 95–6

NATO (North Atlantic Treaty Organization) 109, 110
naturalized geopolitics 12, 26–7, 58, 69, 85, 93–101, 110, 112
natural selection 95–6
navigation 19, 20
Nazi regime 29, 68, 96, 97, 99, 100; anti-Semitism 96, 97, 101
Nehru, Jawaharlal 108
'netwar' 77
neutralism 109
New Zealand: maps of 19
Nicaragua: Contra rebels 106, 111
Nietzsche, Friedrich 15
Ninkovich, Frank 29
Nixon, Richard Milhous 71, 112
Noah (Biblical figure) 23
nomadic empires 72
Non-Aligned Movement 108
non-sovereignty 56
North Atlantic Treaty Organization (NATO) 109, 110
North Korea 2, 4, 77
Norway: independence 99
nuclear threat 31, 58, 77, 105; balance of terror 109
Nussbaum, F. 27

Occidentalism 25
offshore financial centres 64
O' Loughlin, J. 110
open trade see 'market access' economy
oppositional discourses 9
organism: state as 68, 97–8
Orient see East/Orient
Ottoman Empire 44, 92
Ó Tuathail, G. 106, 110, 112

Pagden, A. 24, 38, 52
Parsons, Talcott 46
Pax Americana 76–7
periodization in geopolitical imagination 85–6, 128
perspective 21–23
Philadelphian system 72
Picart, Bernard 91
Pigafetta, Antonio 18–19
Pius II, Pope 23
pluralism 80
Polanyi, Karl 129
political geography 10, 51, 68, 95, 127, 128–3

political theology 69
Porter, Michael 78
Portugal 100
post-colonial studies 58
postmodern world 12, 118–20
power: balance of 24, 67; coercive 55–8; despotic 55, 56; diffusion of 80; infrastructural 55, 56; multipolarity of 75; organization of 73, 77–83; relative power differences 70, 71–2; spatiality of 51–65, 128–32; *see also* Great Powers; superpowers
primacy: future of 81–2; historical geography of 74–81; pursuit of 12, 67–81, 82, 124, 125
Prisoner's Dilemma game 80, *81*
privatization 69
property: Aboriginal land claims 38, 39; mobile 59, 60; rights and statehood 58–60
Ptolemy 6, 17–18, 31, 36

race 26, 37, 48; racist theories 95–7
rationalization 47
Ratzel, Friedrich 6, 68, 96, 99
Reader's Digest 106
Reagan, Ronald 9, 30, 106, 111
regional trading agreements 78
Reich, Robert 74, 79
relative power differences 69, 70–1
Renaissance 16, 17, 18–22, 32, 42, 44
Renan, Ernest 95
Respublica Christiana 23, 44
Ribeiro, Diego 18
Roman Catholic Church 42
Roman Empire: as cultural referent 24, 39, 40, 41, 44–5, 88, 91, 92; influence on Europe 24; and race 26; Renaissance interest in 42
Romanticism 22
Roosevelt, Franklin Delano 107
Rosenberg, E. 104
Rostow, W. 64
Rousseau, Jean-Jacques 52
Ruggie, J.G. 59
Russia 89; Civil War 104; colonialism in 89, 92, 110; *see also* Soviet Union
Ryan, Michael 43

Said, Edward 24, 44
Sandinista regime, Nicaragua 111
Satan 30; Satanic representations 69–70
Schmitter, P. 100

Scotland 88
Scott, Allen 78
sea powers 28
Second World War 101
sedentary tribes 72
September 11, 2001 1, 2, 115, 116, 120, 121, 122
Shapiro, Michael J. 37, 71, 121
Sharp, Joanne 106
Signorelli, A. 41
Simpson, D. 28
Slovenia 7
Smith, A.D. 92
social harmony 98
socialist internationalism 100
'social life' of states 67–9
social sciences 51, 65
society: contained by state 53, 54; hierarchies of 36–7, 42–7; rise of 52
Somoza, Anastasio Somoza Debayle 111
South Africa 80
sovereignty *see* state sovereignty
Soviet Military Power 105
Soviet Union 30; and Cold War 31, 102–12; economy 76, 78, 105; foreign intervention by 108; socialist model 103; US containment of 110–11; as US rival for primacy 76; and Vietnam war 7; *see also* Russia
space: and time 35–48, 99; visualized 11, 15–32
Spain 88, 100
spatiality of power 51–65, 128–32
Spengler, Oswald 98
sporting metaphors 71
Springborg, Patricia 24, 44
Sri Lanka *see* Ceylon
Stalin, Joseph 100, 102, 107
standard international relations theory 54
state: contains society 51, 53; identity 68, 87, 92, 97; as organism 68, 96–7; statehood 10, 11–12, 58–60; *see also* city-states; 'continental' states; 'marchland' states; 'maritime' states; nation-states; territorial states
state-centricity 10, 31–2, 50–1, 75, 122, 129; in early maps 17
state corporatism 100
stateless terrorists 1, 4, 77,121–23
state sovereignty 11, 51, 53–4, 56, 58; and currency 63–4; inequality in 67–9
Stokes, E. 90

superpowers 8, 108; *see also* Soviet Union; United States
supranational currencies 63–4
supranationalism 131
suzerainties 72
Swiss Confederation 56
Syria 77

Taliban 1, 121
Tasman, Abel 19
technological innovations 21, 22, 29, 46, 99
temperate zones 26, 32
temporal metaphor 85
territorial states 11–12, 22, 51–65, 87, 129–30; competition between 12, 67–83; future of 129; hierarchy of 8–10, 22, 29, 67, 68, 85–6, 112; historical geography of 74; internationalization of 79; organization of 72
'territorial trap' 53
terrorists, stateless 1, 4, 77, 121–23
theology: political 69
Thevet (cosmographer) 19
Third World 30, 31, 103; and Cold War 47, 108
'Three Worlds' concept 11, 31, 47, 108
Thucydides 28
time: and global space 35–48, 99; innovations and 21
'time into space' 11
Tito (Josep Broz) 108
Tojo, Hideki 72
Topinambas, Brazil *20*
torrid zones 26, 32
totalization of places 36, 40
totemism 93
Toulmin, S. 86
trade: internationalization of 79–81
transnational liberalism 10, 61–4, 116; hegemony of 61
trasformismo 41
Treaty of Westphalia (1648) 12, 19, 57, 72, 74
tropical zones 26–7
Truman, Harry S 5, 106, 110, 111
Tullio-Altan, Carlo 40–1
Turkey: Ottoman empire 11, 44, 88, 92

UNCTAD (United Nations Conference on Trade and Development) 109
United Nations 52, 76, 109

United States 54, 88; army geography manuals 101; becomes nation-state 98; and Cold War 30–1, 102–12; colonialism 30, 87, 89, 91, 111; and conflict in former Yugoslavia 7; containment doctrines 29–30, 75, 110–11; early organization of 72; exaggeration of Soviet capabilities 105; foreign policy 4, 5, 102–3, 104, 105, 107, 110–12; as hegemonic state 9, 72, 80, 83, 101; and international anarchy axiom 72; and Japan 74; local currencies schemes 64; and 'market access' economy 79–80, 81–2; military power 77–8; model for modernity 8, 10, 46; models from antiquity 88; power of presidency 106, 110; race theories 96; use of sporting metaphors 71–2; Vietnam war 7–8, 29; world economy opened up by Cold War 61; *see also* American Colonies
universe 17; *see also* cosmography

Vietnam War 7–8, 29, 111
'view from nowhere' 11, 15–16, 127
visualization of global space 10–11, 15–33, 85, 121
voluntarism 95
voyages of discovery 17–18, 19, 88

Wallerstein, Immanuel 9
Waltz, Kenneth 54
war: and Christianity 69; information 80; and revolution 86; sporting metaphors 71; *see also* Cold War; First World War; Greek Civil War; Korean War; Second World War; Vietnam War
Warsaw Pact (1955) 109
Weber, Max 46
Weltpolitik 99
Westphalia: Treaty of (1648) 12, 19, 57, 72, 74
Westphalian system 12, 57, 72
Wilhelm II, emperor of Germany and king of Prussia 68
'winner-take-all' markets 73–4
women: symbolize race 27
World Bank 52, 76, 109
world capitalism 59, 87, 118
world culture 82
world economy: 58, 78, 82, 84, 124; breakdown of 100; and economic nationalism 99; and growth of nationalism 93; 'market access' 58–60; origins of 75–6; US hegemon for stability 112

world politics 1, 5, 10, 85, 116, 127; and
ages of geopolitics 85–6; and international
relations 51
World Social Forum (Porto Alegre, Brazil)
116

Yalta Conference (1945) 107
Yugoslavia 7

Zapatista rebels 77

ESSENTIAL READING

Political Geography

Mark Blacksell

Hb: 0-415-24667-9
Pb: 0-416-24668-7   Routledge

An Introduction to Political Geography: Space, Place & Politics

Martin Jones, Rhys Jones & Michael Woods

Hb: 0-415-25076-5
Pb: 0-415-25077-3   Routledge

The Geopolitics Reader

Gearoid O.Tuathail, Simon Dalby & Paul Routledge

Hb: 0-415-16270-X
Pb: 0-415-16271-8   Routledge

Geopolitical Traditions
Critical Histories of a Century of Geopolitical Thought
Klaus Dodds & David Atkinson

Hb: 0-415-17248-9
Pb: 0-415-17249-7   Routledge

Information and ordering details

For price availability and ordering visit our website www.tandf.co.uk

Subject Web Address   **www.geographyarena.com**

Alternatively our books are available from all good bookshops.